An Atlas of Airborne Pollen Grains and Common Fungus Spores of Canada

An Atlas of Airborne Pollen Grains and Common Fungus Spores of Canada

I. John Bassett, Clifford W. Crompton,
and John A. Parmelee

Biosystematics Research Institute
Ottawa, Ontario

Research Branch
Canada Department of Agriculture
Monograph No. 18
1978

© Minister of Supply and Services Canada 1978

Available by mail from

Printing and Publishing
Supply and Services Canada
Hull, Québec, Canada K1A 0S9

or through your bookseller.

Catalogue No. A54-3/18　　　　　　　　　　　Canada: $12.00
ISBN 0-660-10016-9　　　　　　　Other countries:　14.40 (Canadian Funds)

Price subject to change without notice.

CODE No. 3M-39078-9:78
Thorn Press Limited
CONTRACT NO. 09KT 01A05-8-39078

CONTENTS

Introduction	9
Part 1 Airborne Pollen Grains	11
Ragweed Pollen Air Indexes for Canada	16
Materials, Methods, and Techniques Used for Pollen Descriptions	40
Key to the Pollen Classes	41
Keys to the Pollen and Spores of 145 Canadian Taxa	43
Descriptions of Nonfungus Spores and Pollen Grains of Vascular Plants	48
Fern allies	48
Gymnosperms	50
Angiosperms	96
Glossary	259
References	263
Part 2 Common Airborne Fungus Spores	269
Descriptions of Fungus Spores	275
Glossary	313
References	315
Acknowledgments	317
Index of Species Described	319
Vascular Plants	319
Fungi	320

INTRODUCTION

Many workers in allergology, botany, entomology, palynology, and other disciplines are interested in identifying airborne pollen grains and fungus and nonfungus spores.

Some information is available on airborne pollen and spores for parts of Canada, but little detailed information has been published for the whole country.

Descriptions and keys of 143 pollen and several spore types that have been found to be airborne in different locations are presented. These are accompanied by photographs made by using light and scanning electron microscopy. The general abundance, distribution of taxa, time of pollen shedding, significance in causing hay fever, and other pertinent data are included.

Many palynologists have devised different terminology to describe pollen grain structure and sculpturing. The systems of terminology of Faegri and Iversen (1964), Erdtman (1966), Van Campo-Duplan (1954), Kuprianova (1956), and Larson et al. (1962) were assessed and it seemed practical to use mainly the terminology of Faegri and Iversen.

The taxonomic treatment of names conforms to the monograph of Payne (1964), and the following books: *Native Trees of Canada* by Hosie (1969), *Checklist of Native and Naturalized Trees of the United States* by Little (1953), *Weeds of Canada* by Frankton and Mulligan (1970), *Manual of the Vascular Plants of Northeastern United States and Adjacent Canada* by Gleason and Cronquist (1963), and *Vascular Plants of the Pacific Northwest* by Hitchcock et al. (1961-1964).

Approximately 20% of all Canadians are allergic to airborne pollen, fungus or mold spores, algae, mites, animals, cloth, dust, fumes or other irritants. This publication presents information on airborne pollen and a number of fungus spores, which are the main causes of hay fever and other forms of allergy.

The following general publications contain descriptions of pollen and spores from North and South America, Europe, and Asia: Erdtman (1957, 1966, 1969); Erdtman et al. (1961, 1963); Faegri and Iversen (1964); Heusser (1971); Hyde and Adams (1958); Kapp (1969); Martin and Drew (1969, 1970); McAndrews et al. (1973); Praglowski (1962); Richard (1970); and Wodehouse (1935, 1971).

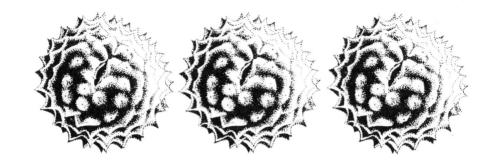

Part 1 AIRBORNE POLLEN GRAINS

FUNCTION AND STRUCTURE OF POLLEN

Pollen grains are extremely small reproductive structures contained in the anthers, or male elements, of seed plants. They are transferred to the female reproductive body mainly by wind, water, or insects.

Dispersal by wind is a haphazard and inefficient method of pollination and it sometimes requires a vast supply of pollen. Some wind-pollinated plants have flowers with many stamens and as many as 10 000 pollen grains per stamen. Only a minute amount of pollen completes its function in the reproductive cycle of the plant. Pollen that does not happen to land on the female organ (stigma) of a flower eventually falls and becomes a part of the accumulation of sediments on the earth's surface.

Pollen grains have many different shapes and external features. Because the outer layer, or exine, of most pollen is not easily dissolved, these grains have been preserved without damage in peats and sediments throughout the geological ages.

Fresh pollen grains can be treated chemically so that only the exine remains and this outer shell wall can be compared with fossil pollen. These fossil remains are important in tracing the history of plants throughout the geological ages (Bassett 1965).

COLLECTION OF AIRBORNE POLLEN

For collecting the airborne pollen, a device called the gravity or Durham sampler was used (Fig. 1). It was approved by the Committee on National Pollen Survey of the American Academy of Allergy in 1946. The procedure consisted essentially of a 24-h exposure of glycerine jelly coated slides in a standard air-sampling device. The ragweed pollen over a unit area of 1 cm^2 was counted and converted into approximate volumetric equivalents by using standard factors. For example, the number of pollen grains of common ragweed, *Ambrosia artemisiifolia* L., per square centimetre was multiplied by 3.6 to obtain the volumetric count. The devices were placed well above the ground to permit complete air movement on all sides. The intensity and duration of the ragweed pollen season at any particular site can be expressed by a single figure (Durham 1937). This air index was derived from the formula $\frac{A}{200} + \frac{B}{100} + C$, where A is the total number of ragweed pollen grains per 0.76 m^3 (cu yd) of air for the ragweed season, B is the total pollen per 0.76 m^3 of air for the peak day, and C is the number of days with 25 or more pollen grains per 0.76 m^3 of air.

Fig. 1. Gravity or Durham pollen sampler.

Individuals sensitive to ragweed pollen are usually sensitive to pollen grains of poverty weed, *Iva axillaris* Pursh; false ragweed, *I. xanthifolia* Nutt.; and cockleburs, *Xanthium* spp. Pollen grains of these plants on test slides were added to those of the ragweeds in computing the air indexes (Bassett and Frankton 1971).

Samter and Durham (1955) prepared a comprehensive and practical survey on a broader scale of the local allergic problems of the United States and its neighboring countries, Canada, Mexico, and Cuba.

Since the advent of the gravity or Durham apparatus there has been a continuous improvement in airborne pollen sampling techniques. Smith and Rooks (1954) studied hourly variations of ragweed pollen at Iowa City, Iowa, during August and September 1952 by using a continuous recording volumetric particle sampler previously described by Stenburg and Hall (1956). Hyde (1959) obtained much more airborne pollen data in South Wales with the Hirst automatic volumetric spore trap (Hirst 1952) than with the gravity type of apparatus. P. W. Voisey of the Engineering Research Service, Canada Department of Agriculture, Ottawa, designed a new continuous volumetric pollen sampler, which includes the desirable features of several others (Fig. 2). This apparatus has been described by Voisey and Bassett (1961) and the results obtained at Ottawa in late August and September 1960 have been included. The peak pollen period was mainly between 7 and 10 a.m. After this peak period there was a decline during the rest of the day; little ragweed pollen was collected at night. From volumetric counts taken hourly during the 1961 ragweed season at Ottawa (Holmes and Bassett 1963), it was found that periods of maximum airborne pollen concentration were directly related to maximum air turbulence, for example, development of superadiabatic lapse. In many cases a sharp decrease in relative humidity decreased several hours in advance of the development of a superadiabatic lapse with no increase in pollen count. A nocturnal rise in pollen count was often noted in calm, moderately humid air with neutral to slightly lapse rates. This may be the result of reflotation or of the breakup of a nocturnal inversion that carries pollen-laden air to the surface.

The intermittent rotoslide sampler (Raynor and Ogden 1970; Ogden et al. 1974) and other similar devices have been in use at several sites in the United States and other countries for the last few years. These devices are much more efficient than the gravity or Durham sampler and may play a much more important role in the future in carrying out airborne pollen studies.

AIRBORNE RAGWEED COLLECTIONS AT OTTAWA USING THE GRAVITY SAMPLER, 1952–1974

The gravity or Durham sampler was placed on the roof of a building situated near the center of the 607 ha at the Central Experimental Farm, Ottawa. Slides were exposed for 24-h periods during August and September from 1952 to 1974. Figure 3 shows two comparative graphs with average ragweed pollen counts for each day from 1952 to 1961 and from 1962 to 1971. The peak ragweed pollen counts on both graphs occur roughly between August 15 and September 15. At stations in southern Ontario and Quebec over a 2- or 3-yr period the same peak counts occurred as at Ottawa. Figure 4 shows the ragweed pollen air indexes at Ottawa between 1953 and 1974. Overall counts were higher between 1952 and 1961 than between 1962 and 1974. There was a marked reduction in the ragweed pollen air indexes at Ottawa from 1968 to 1974. All indexes in this period were below 10. The index figures can be attributed partly to the changes of crops produced in eastern Ontario, particularly in Carleton County where seed corn has accounted for over 40% of the cereal acreage in the last few years (Ontario Ministry of Agriculture and Food 1973). Gebben (1965) has pointed out the greater abundance of ragweed in cereal crops than in row crops such as corn. This difference is related primarily to the nature of agricultural practices affecting these crops. Cultivation of row crops with modern equipment kills many weed seedlings including ragweeds, and chemical sprays are more effective in controlling these plants in corn than in other cereal crops. Rotation practices also help to reduce the abundance of ragweed in one given area.

Fig. 2. Continuous volumetric pollen sampler.

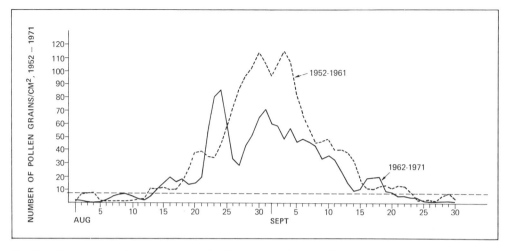

Fig. 3. Comparison of average ragweed pollen counts for each day, 1952–1961; 1962–1971.

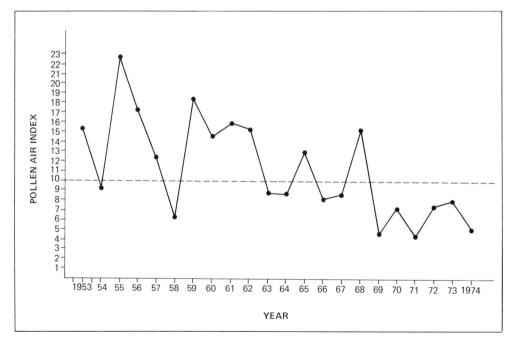

Fig. 4. Ragweed pollen air indexes at Ottawa, 1953–1974.

Another factor that contributes to the reduction of ragweed pollen is the suburbanization of the area around the Central Experimental Farm in Ottawa. Solomon and Buell (1969) pointed out that pollen counts in 1967 at Yonkers, N.Y., were less than half the counts conducted in 1932. Decreases occurred in all herb taxa except *Artemisia* (sage); grasses and ragweeds demonstrated the greatest decline. This change seems to be related to the massive suburbanization to the north and west of the sampling site after World War II and the subsequent massive decrease in sites available for the growth of wind-pollinated herbs.

Further information on the biology of the common and perennial ragweeds in Canada was published recently (Bassett and Crompton 1975).

RAGWEED POLLEN AIR INDEXES FOR CANADA

Allergists have classified the suitability of communities for hay fever sufferers according to the ragweed pollen air index.

Above 10 is not recommended for ragweed hay fever sufferers.
Between 5 and 10 is fairly good.
Below 5 is good.
Below 1 is excellent.

Province and locality	Period of ragweed pollen survey	Average ragweed pollen air index
British Columbia		
Cranbrook	1965	0.0
Field (Yoho National Park)	1965	0.0
Glacier (Glacier National Park)	1965	0.0
Crescent Valley	1965	0.1
Creston	1965	0.1
Grand Forks	1965	0.1
Summerland	1950	0.0
	1965	0.0
Vernon	1965	0.0
Kamloops	1965	0.0
Hope	1965	0.0
Williams Lake	1965	0.0
Prince George	1965	0.0
Vancouver	1965	0.0
Saanichton	1953–54	0.0
	1965	0.1

Province and locality	Period of ragweed pollen survey	Average ragweed pollen air index
Victoria	1958	0.8
Parksville	1965	0.1
Comox	1965	0.0
Queen Charlotte City	1965	0.3
Alberta		
Cypress Hills	1952	0.0
Manyberries	1950–51	0.1
Medicine Hat	1950–51	7.0
Vermilion	1950–51	0.0
Lethbridge	1950–51	1.0
Drumheller	1950–51	1.0
Edmonton	1950–51	0.0
Waterton Lakes National Park	1952	0.0
Coleman	1950–51	0.0
Calgary	1950–51	0.0
Banff (Banff National Park)	1950–51	0.0
Mt. Eisenhower (Banff National Park)	1968	0.0
Lake Louise (Banff National Park)	1950–51	0.0
Jasper (Jasper National Park)	1952	0.0
Beaverlodge	1950–51	0.0
Vegreville	1968	0.0
Saskatchewan		
Carlyle	1960–62	1.6
Estevan	1960–62	2.5
Weyburn	1960–62	2.9
Regina	1955	0.3
Melfort	1955	0.1
Saskatoon	1951–53	0.3
Prince Albert	1930	0.1
Prince Albert National Park (Waskesiu)	1951	0.0
Swift Current	1952–53	1.3
Scott	1955	0.1
Maple Creek	1968	0.0
Govenlock	1968	0.0
Manitoba		
Winnipeg	1947–54	7.0
	1960–62	4.1
Steinbach	1960–62	2.1
Morris	1960–62	18.6

Province and locality	Period of ragweed pollen survey	Average ragweed pollen air index
Manitoba (cont'd)		
Emerson	1960–61	4.7
Morden	1940	12.0
	1960–62	3.6
Mather	1960–61	2.3
Pierson	1940	6.0
	1960–62	3.5
Brandon	1961–62	5.0
Portage la Prairie	1960–62	1.5
Russell	1940	1.0
Riding Mountain National Park	1950	0.2
Dauphin	1940	5.0
The Pas	1940	0.1
Ontario		
Cornwall	1953–54	22.2
Ottawa	1950–74	10.0
St. Lawrence Islands National Park (Mallorytown)	1950–52	33.2
Smiths Falls	1957–59	14.2
Westport (Rideau Lakes)	1957–59	7.6
Kingston	1961–73	29.4
Calabogie	1963–64	9.2
Renfrew	1958–60	4.9
Picton	1956	38.2
Belleville	1956	30.2
Madoc	1957–59	21.4
Presqu'ile Park	1963	18.4
Kasshabog Lake (near Marmora)	1963–64	9.6
Pembroke	1958–60	4.5
Barry's Bay	1957–59	1.1
	1963–64	7.5
Bancroft	1955–57	8.1
Chalk River	1954–56	4.5
Peterborough	1953–54	33.4
Haliburton	1956–58	1.9
	1963–64	3.8
Mattawa	1958–60	1.5
Algonquin Park	1952–55	12.3
	1963	1.5
Dorset	1952–54	6.1
Muskoka Falls	1955–57	4.6
Gravenhurst	1955–57	16.8

Province and locality	Period of ragweed pollen survey	Average ragweed pollen air index
Ontario (cont'd)		
Huntsville	1953–56	9.4
Toronto area:		
Core of Central Zone	1957–67	35.6
Southwest Metro	1965–67	33.6
Metro Central	1957–67	42.5
East Metro	1957–67	44.6
Northwest Metro	1957–67	54.0
Midland	1954	11.5
Georgian Bay Islands National Park		
(Beausoleil Island)	1950–53	14.8
Port Carling	1955–57	9.6
Rosseau	1957–59	3.7
Lake Joseph (Muskoka)	1951	4.0
Parry Sound	1955–56	19.4
Magnetawan	1957–59	3.5
	1963–64	6.0
South River	1957–59	1.8
North Bay	1951–53	7.5
Temagami	1954–55	2.4
New Liskeard	1956–57	0.3
Timmins	1958–59	0.2
Cochrane	1934–35	2.0
Kapuskasing	1951–52	0.4
Sudbury	1954–55	3.4
Hamilton	1961–68	17.2
Guelph	1963–64	30.5
London	1953–54	38.5
Point Pelee National Park	1950–52	38.5
Port Franks	1964	37.5
Kincardine	1958–60	22.4
Wiarton	1958–60	16.9
Inverhuron Park	1964	27.4
Lion's Head	1958–60	18.3
	1963–64	11.4
Tobermory	1956–58	5.2
Mindemoya (Manitoulin Island)	1952–55	7.7
Espanola	1956–59	3.9
Blind River	1956–57	2.5
Sault Ste. Marie	1952–54	6.2
Thunder Bay	1957–59	0.9
10 miles southwest of Thunder Bay	1956	0.1
Black Sturgeon Lake		
(Thunder Bay District)	1952	2.3

Province and locality	Period of ragweed pollen survey	Average ragweed pollen air index
Ontario (cont'd)		
Fort Frances	1956–57	1.0
Cedar Lake (Kenora District)	1952–54	3.4
Kenora	1956–58	6.2
Quebec		
Matapédia	1938	0.1
Carleton	1949–56	0.7
New Carlisle	1938	3.0
Chandler	1938	0.1
Grand Rivière	1949–52	0.2
Percé	1949–56	0.7
Gaspé	1949–56	0.2
Îles-de-la-Madeleine	1941	0.1
Mont-Albert	1939	0.1
Matane	1954–56	2.2
Mont-Joli	1938	0.2
Father Point	1934–35	1.0
Rimouski	1949–56	3.0
Rivière-du-Loup	1949–56	4.3
Tadoussac	1966–70	0.3
Jonquière (Chicoutimi)	1953–55	3.0
Normandin	1939–41	3.0
Baie St. Paul	1966–70	12.9
Ste. Anne-de-la-Pocatière	1967	9.5
Charlesbourg	1939–41	2.0
Québec City	1949–55	11.1
Sherbrooke	1951–55	16.4
Lennoxville	1939–41	4.0
Victoriaville	1951–55	29.6
Cap-de-la-Madeleine	1953–55	43.4
Berthierville	1939–41	33.0
St.-Laurent-d'Orléans	1967–70	17.2
Mont-Orford	1966–71	11.5
Farnham	1939	64.0
Montreal area:		
Dorval	1962–67	48.4
McGill University	1962–63	25.1
Beaconsfield	1961–67	22.6
Ste. Anne de Bellevue	1950–55	37.9
St. Jerome	1960–71	18.2
Lac-des-Seize-Îles	1949–52	9.1
Mont-Roland	1965	2.1
Ste. Marguerite	1964–65	5.7

Province and locality	Period of ragweed pollen survey	Average ragweed pollen air index
Quebec (cont'd)		
Ste. Agathe	1952–65	6.7
St. Adolphe	1964–71	2.3
St. Jovite	1952–64	3.1
St. Faustin	1967–71	2.9
Mont-Tremblant	1952–64	3.5
Labelle	1960–64	3.8
Nominingue	1952–56	6.5
Lac-des-Plages	1956–64	6.2
Mont-Laurier	1953–55	5.2
Luskville	1950–51	21.0
New Brunswick		
Sackville	1952–68	0.4
Pointe du Chene	1952–53	20.0
	1954–68	1.5*
Shediac Cape	1952–54	0.6
Moncton	1952–68	0.2
Fundy National Park	1950–55	5.5
	1959–68	0.3*
Sussex	1952–68	0.4
Chipman	1952–68	0.4
Jemseg	1952–67	2.2
Gagetown	1952–53	19.5
	1954–68	0.7*
Saint John	1952–68	0.3
Welsford	1952–68	0.3
Fredericton	1950–68	0.1
St. George	1952–68	0.3
St. Andrews	1952–68	0.3
St. Stephen	1952–68	0.4
Grand Manan	1956–68	0.3
McAdam	1956–68	0.4
Woodstock	1952–68	0.1
Perth-Andover	1952–68	0.1
Edmundston	1952–68	0.9
Doaktown	1952–68	0.2
Richibucto	1958–68	0.2
Newcastle-Chatham	1952–68	0.1
Tracadie	1956–67	0.7
Bathurst	1954–68	0.1

*This figure is included to indicate the reduction in the pollen air index since the inception of control campaigns in 1954.

Province and locality	Period of ragweed pollen survey	Average ragweed pollen air index
New Brunswick (cont'd)		
Dalhousie	1952–68	0.1
Campbellton	1952–68	0.0
Charlo	1967	0.0
Prince Edward Island		
Souris	1952–56	1.0
Montague	1952–56	0.6
Charlottetown	1952–56	1.4
P.E.I. National Park (Dalvay House)	1952–56	3.0
Summerside	1952–56	1.0
O'Leary	1952–56	1.4
Tignish	1952–56	1.2
Nova Scotia		
Ingonish Island	1950–55	1.4
Cape Breton Highlands National Park (Ingonish Beach)	1950–55	0.9
Baddeck	1951–54	0.4
Sydney	1967	0.3
Antigonish	1951–55	0.4
Truro	1950–54	0.2
Kentville	1953–55	4.7
Halifax	1954	1.9
Chester	1951–55	0.3
Digby	1951–55	3.2
Meteghan	1951–52	4.5
Yarmouth	1952–56	4.5
Middle West Pubnico	1951	0.3
Sable Island	1967	0.2
Newfoundland		
St. John's	1950–55	0.3
Corner Brook	1951 and 1955	0.2
Mount Pearl (near St. John's)	1954	0.1
Gander	1967	0.1

Figure 5 shows where the heaviest concentrations of ragweed airborne pollen grains have been found in Canada.

Table 1 lists all the hay-fever plants, including trees and shrubs, known in Canada and their flowering periods, distribution, and abundance.

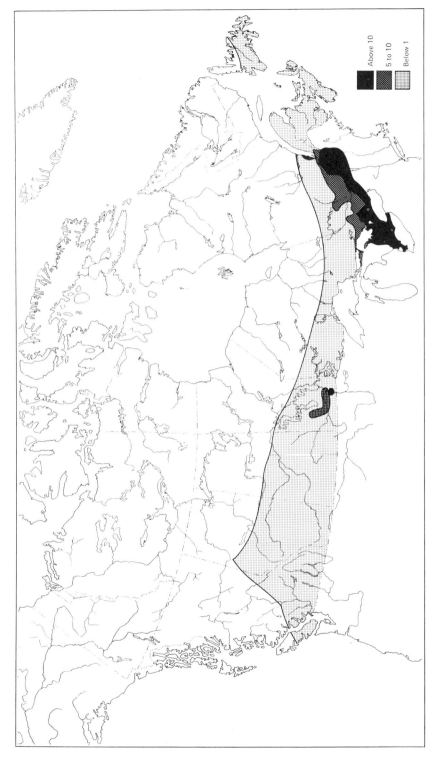

Fig. 5. Map of Canada showing the distribution of ragweed pollen air indexes.

Table 1. Flowering period and relative abundance of hay-fever plants

	Flowering period	B.C.	Alta.	Sask.	Man.	N.Ont
PTERIDOPHYTA (Vascular cryptogams)						
Club-moss (*Lycopodium selago*)	July–Sept.	+++	++	++	++	++++
SPERMATOPHYTA (Seed plants, flowering plants)						
Trees and Shrubs						
Alder, green (*Alnus crispa*)	June–July		++	++	++	+++
Alder, speckled (*Alnus rugosa*)	Mar.–June			+	+	+++
Ash, black (*Fraxinus nigra*)	May–June					++
Ash, white (*Fraxinus americana*)	May–June					++
Ash, red (*Fraxinus pennsylvanica*)	May–June					+
Aspen, largetooth (*Populus grandidentata*)	May				+	++
Basswood (*Tilia americana*)	June–July				+	++
Bayberry (*Myrica pensylvanica*)	May–July					+
Beech (*Fagus grandifolia*)	Apr.–May					++
Birch, dwarf white (*Betula minor*)	June–July					
Birch, gray (*Betula populifolia*)	Apr.–May					

symbols: + rare; ++ occasional; +++ common; ++++ abundant.
The ratings were based on field surveys, herbarium collections, and recent botanical literature.

Ont.	N.Que.	S.Que.	N.B.	P.E.I.	N.S.	Nfld.	Y.T.	N.W.T.
++	++++	++	+++	++	+++	++	+++	+++
+++	+++	++	+++	++	+++	+++	++	++
++	+++	+++	+++	++	+++	+++		
++	++	++	++	++	++			
++	++	+++	+++	+++	+++			
++	+	++	++	++	++			
++	++	+++	+++	++	++			
++	++	+++	++					
++	+	++	++	++	++	++		
++	++	+++	+++	+++	+++			
	+++					+++		
	++	+++	+++	++	++			

Table 1. Flowering period and relative abundance of hay-fever plants (cont'd)

	Flowering period	B.C.	Alta.	Sask.	Man.	N.O.
Birch, water (*Betula occidentalis*)	May–June	+++	+++	+++	+++	+
Birch, white (*Betula papyrifera*)	Apr.–May	++++	+++	++	+++	+++
Birch, yellow (*Betula alleghaniensis*)	Apr.–May					+
Butternut (*Juglans cinerea*)	June					
Cedar, eastern white (*Thuja occidentalis*)	May				++	+
Cedar, western red (*Thuja plicata*)	Apr.–May	+++	+			
Chestnut (*Castanea dentata*)	June–July					
Cypress, yellow (*Chamaecyparis nootkatensis*)	Apr.	+++				
Elm, slippery (*Ulmus rubra*)	Apr.–May					
Elm, white (*Ulmus americana*)	Apr.–May			+	++	+
Hackberry (*Celtis occidentalis*)	Apr.–May					
Hazel, beaked (*Corylus cornuta*)	Apr.	++	++	++	++	+
Hemlock, eastern (*Tsuga canadensis*)	May					+
Hemlock, mountain (*Tsuga mertensiana*)	Apr.–May	++				
Hemlock, western (*Tsuga heterophylla*)	Apr.–May	++				
Hickory, bitternut (*Carya cordiformis*)	May–June					

Ont.	N.Que.	S.Que.	N.B.	P.E.I.	N.S.	Nfld.	Y.T.	N.W.T.
+++	++++	++++	++++	+++	++++	++++	+++	+++
+++	++	+++	+++	+++	+++	++		
++	+	++	++					
+++	++	+++	+++	++	++			
++		++						
++		++						
+++	++	+++	+++	++	+++	+		
++		++						
++	++	++	++	+	++	+		
++	++	++	++	++	++			
++	+	++						

Table 1. Flowering period and relative abundance of hay-fever plants (cont'd)

	Flowering period	B.C.	Alta.	Sask.	Man.	N.Ont
Hickory, pignut (*Carya glabra*)	May–June					
Hickory, shagbark (*Carya ovata*)	May–June					
Hop-hornbean (*Ostrya virginiana*)	Apr.–May				+	+
Juniper, common (*Juniperus communis*)	Apr.–May	+++	+++	++	++	+++
Maple, bigleaf (*Acer macrophyllum*)	Apr.–May	+++				
Maple, black (*Acer nigrum*)	Apr.–May					
Maple, Douglas (*Acer glabrum* var. *douglasii*)	Apr.–May	+++	++			
Maple, Manitoba (*Acer negundo*)	May–June	++	++	+++	++++	++
Maple, mountain (*Acer spicatum*)	May–June			++	++	+++
Maple, red (*Acer rubrum*)	Apr.–May					++
Maple, silver (*Acer saccharinum*)	Apr.–May					++
Maple, striped (*Acer pensylvanicum*)	May–June					++
Maple, sugar (*Acer saccharum*)	Apr.–May					+++
Maple, vine (*Acer circinatum*)	Apr.–May	++				
Mulberry, red (*Morus rubra*)	May–June					
Oak, black (*Quercus velutina*)	May–June					

Ont.	N.Que.	S.Que.	N.B.	P.E.I.	N.S.	Nfld.	Y.T.	N.W.T.
++		++						
++		++						
++	+	++	++	++	++			
+++	+++	+++	+++	++	+++	+++		
++		+						
+++		++						
+++	++	+++	+++	++	++	++		
+++	++	+++	+++	++	+++	+++		
+++	++	+++	++					
++	++	+++	+++	++	+++			
+++	+++	++++	++++	+++	+++			
++								
++								

Table 1. Flowering period and relative abundance of hay-fever plants (cont'd)

	Flowering period	B.C.	Alta.	Sask.	Man.	N.On
Oak, bur (*Quercus macrocarpa*)	June				+	++
Oak, Chinquapin (*Quercus muehlenbergii*)	June					
Oak, garry (*Quercus garryana*)	June–July	++				
Oak, pin (*Quercus palustris*)	Apr.–May					
Oak, red (*Quercus rubra*)	May–June					++
Oak, swamp white (*Quercus bicolor*)	June					
Oak, white (*Quercus alba*)	May					
Pine, jack (*Pinus banksiana*)	May	+	++	+++	+++	++++
Sweet fern (*Comptonia peregrina*)	Apr.–June					++
Sweet gale (*Myrica gale*)	Apr.–June	++	++	+	+	++
Sycamore (*Platanus occidentalis*)	May					
Walnut, black (*Juglans nigra*)	June					
Willow (*Salix* spp.)	Apr.–May	+++	+++	++	++	+++
Grasses (Gramineae)						
Corn (*Zea mays*)	July–Aug.	+++	++	+	+	+
Grass, quack (*Agropyron repens*)	June–Aug.	+++	+++	++	++	++

.Ont.	N.Que.	S.Que.	N.B.	P.E.I.	N.S.	Nfld.	Y.T.	N.W.T.
+++	+	+++	++					
++								
++								
+++	++	+++	+++	++	+++			
+++		++						
+++	++	++						
+++	++++	+++	++	++	+++	+++	+++	+++
+++	++	+++	+++	++	+++			
+++	+++	+++	+++	++	+++	+++		
++								
++								
+++	+++	+++	+++	+++	+++	+++	++	
+++	+	+++	++	++	+++			
+++	+++	++++	+++	+++	+++	+++	++	++

Table. 1. Flowering period and relative abundance of hay-fever plants (cont'd)

	Flowering period	B.C.	Alta.	Sask.	Man.	N.Ont
Grass, Kentucky blue (*Poa pratensis*)	May–July	++++	+++	+++	+++	++
Grass, orchard (*Dactylis glomerata*)	June–July	+++	++	++	++	+
Grass, timothy (*Phleum pratense*)	June–July	+++	+++	+++	+++	+++
Ragweeds and Their Close Relatives						
Cocklebur, common (*Xanthium strumarium*)	July–Aug.	++	++	++	++	+
Cocklebur, spiny (*Xanthium spinosum*)	July–Aug.					
Beach sandbur (*Ambrosia chamissonis*)	May–Sept.	++				
Elder, marsh (*Iva frutescens*)	July–Aug.					
Poverty weed (*Iva axillaris*)	June–Aug.	++	+++	+++	++	
Ragweed, common (*Ambrosia artemisiifolia*)	July–Sept.	+	++	++	++	+
Ragweed, bur (*Ambrosia acanthicarpa*)	July–Sept.		++	++		
Ragweed, false (*Iva xanthifolia*)	July–Sept.	++	+++	+++	+++	
Ragweed, giant (*Ambrosia trifida*)	July–Sept.	+	+	+	++	+
Ragweed, perennial (*Ambrosia psilostachya*)	July–Sept.	+	+	+++	+++	+
Sage (*Artemisia* spp.)						
Mugwort (*Artemisia vulgaris*)	Aug.–Sept.	++	+	+	+	+

	Ont.	N.Que.	S.Que.	N.B.	P.E.I.	N.S.	Nfld.	Y.T.	N.W.T.
	+++	+++	++++	+++	++++	++++	+++	+	+
	+++	++	++++	+++	++	+++	+	+	
	+++	+++	++++	+++	+++	+++	+++	++	+
	+++	++	+++	++	++	++	++		
	++								
						+			
	+++	++	++++	++	+	++	+		+
	+		+	+	+				
	++	++	++	++	++	++			
	+	+			+	+			
	++	++	++	++	++	++	++		

Table. 1. Flowering period and relative abundance of hay-fever plants (cont'd)

	Flowering period	B.C.	Alta.	Sask.	Man.	N.O.
Sage, pasture (*Artemisia frigida*)	Aug.–Sept.	+++	++++	++++	++++	–
Sage, prairie (*Artemisia ludoviciana*)	Aug.–Sept.	+++	++++	++++	++++	–
Sagebrush, big (*Artemisia tridentata*)	Aug.–Oct.	+++	++			
Sagewort, field (*Artemisia campestris*)	July–Aug.	++	+	+	+	+
Wormwood, biennial (*Artemisia biennis*)	Aug.–Sept.	++	+	+	++	+
Miscellaneous Plants						
Cattail, broad-leaved (*Typha latifolia*)	May–July	+++	++	++	+++	++
Cattail, narrow-leaved (*Typha angustifolia*)	May–July				+	
Dock, broad-leaved (*Rumex obtusifolius*)	July–Aug.	+				
Dock, water (*Rumex orbiculatus*)	July–Aug.		+	+	+	+
Kochia (*Kochia scoparia*)	July–Sept.	+++	+++	+++	+++	
Greasewood (*Sarcobatus vermiculatus*)	July–Oct.	++	++	++		
Lambs'-quarters (*Chenopodium album*)	July–Sept.	++++	+++	+++	++++	++
Lambs'-quarters net-seeded (*Chenopodium berlandieri* ssp. *zschackei*)	July–Sept.	++	+++	+++	+++	
Littorella, American (*Littorella americana*)	June–Aug.					
Marijuana (*Cannabis sativa*)	June–Sept.	+	+		+	

Ont.	N.Que.	S.Que.	N.B.	P.E.I.	N.S.	Nfld.	Y.T.	N.W.T.
+	+	+	+					
+	+	+						
++	++	++	+	+	+	+		
++	++	++	++	++	++	++		
++	+++	+++	+++	++	+++	+++	++	++
++	+	++	+	+	+			
++	+	++	++	+	+	+		
++								
+	+	+		+	+			
++	+++	++++	+++	+++	+++	+++	++	++
+								
+		+	+		+++	++		
++	+	++	+	+	+			

Table 1. Flowering period and relative abundance of hay-fever plants (cont'd)

	Flowering period	B.C.	Alta.	Sask.	Man.	N.On
Nettle, false (*Boehmeria cylindrica*)	July–Aug.					
Nettle, stinging (*Urtica dioica ssp. gracilis*)	May–Oct.	+++	+++	+++	+++	++
Nettle, wood (*Laportea canadensis*)	July–Sept.			+	+	+
Orach (*Atriplex subspicata*)	July–Sept.	++	+++	+++	+++	+
Pellitory (*Parietaria pensylvanica*)	June–Aug.	+	+	+	+	+
Pigweed, tumble (*Amaranthus albus*)	July–Aug.	+++	+++	+++	+++	++
Pigweed, redroot (*Amaranthus retroflexus*)	July–Aug.	+++	++	++	++	++
Pigweed, Russian (*Axyris amaranthoides*)	July–Aug.	+	+++	+++	+++	+
Plantain, Alaska (*Plantago macrocarpa*)	May–June	++				
Plantain, common (*Plantago major*)	June–Oct.	+++	+++	+++	+++	+++
Plantain, English (*Plantago lanceolata*)	June–Oct.	++++				+
Plantain, Rugels' (*Plantago rugelii*)	June–Oct.					+
Plantain, saline (*Plantago eriopoda*)	May–July	++	++	++	++	
Plantain, seaside (*Plantago maritima*)	May–July	+++			+	++
Plantain, Siberian (*Plantago canescens*)	May–July	+				

Ont.	N.Que.	S.Que.	N.B.	P.E.I.	N.S.	Nfld.	Y.T.	N.W.T.
++	++	++						
++	++	+++	++	++	++	++	++	+
++	+	++	++	++	++	++		
	++	+	++	+++	+++	++		+
++	+	++						
++	++	+++	+++	+++	+++	++		
++	++	+++	+++	+++	++			
+								
++	+++	++++	+++	+++	+++	+++	+	+
++	+	+++	++	++	++++	+++		
++	+	+++	+	+	+			
	++						+	+
	++		+++	+++	+++	+++	++	+
							++	++

Table 1. Flowering period and relative abundance of hay-fever plants (concl'd)

	Flowering period	B.C.	Alta.	Sask.	Man.	N.Ont.
Richweed (*Pilea pumila*)	July–Oct.					
Russian thistle (*Salsola pestifer*)	July–Sept.	+++	++++	++++	++++	++
Sea-blite (*Suaeda maritima*)	July–Sept.					
Sedge (*Carex aquatilis*)	June–Sept.	+++	++	++	++	+++
Sorrel, garden (*Rumex acetosa*)	June–Aug.	+	+	+	+	+
Sorrel, mountain (*Oxyria digyna*)	May–Aug.	++	++			
Sorrel, sheep (*Rumex acetosella*)	June–Aug.	+++	+++	++	++	+++
Winter fat (*Eurotia lanata*)	July–Sept.		++	++		

.Ont.	N.Que.	S.Que.	N.B.	P.E.I.	N.S.	Nfld.	Y.T.	N.W.T.
+++	+	+++	++	+				
+	+	+	+	+	+			
	+	++	+++	+++	+++	+++		
+++	+++	+++	++	++	++	++	++	++
+	+	++	++	+	+	+		
+++	+++	+++	+++	++	+++	+++	+	

MATERIALS, METHODS, AND TECHNIQUES USED FOR POLLEN DESCRIPTIONS

Pollen grains were taken from dried herbarium specimens in the Vascular Plant Herbarium, Canada Department of Agriculture, Ottawa, and the National Museum of Natural Sciences, Ottawa. They were acetolyzed and mounted in silicone oil according to the method outlined by Bassett and Crompton (1968). This procedure is slightly different from Erdtman's (1966) technique.

Where pollen grains were scarce, individual anthers were crushed in a depressed culture slide and gently boiled in several drops of a mixture of acetic anhydride and sulfuric acid (9:1). The reaction was observed microscopically (40 ×) and stopped by adding several drops of absolute ethyl alcohol after the pollen contents were digested and the exine appeared golden brown to brown in color. After the reaction was stopped, grains were sucked up in a micropipette and deposited on a microscope slide on a small amount of molten glycerine jelly. A cover slip was placed on top and sealed to the slide to exclude the air.

The average size of each pollen type was based on the measurements of at least 20 grains mounted in silicone oil. To confirm these observations the pollen from each taxon from at least five different locations encompassing the geographical range of that species in Canada was examined in glycerine jelly.

Indexes for polar areas were calculated using the ratio of the polar area measurement to the width as outlined by Kremp 1965).

MICROSCOPY

Pollen grains were examined using light, phase contrast, and interference microscopy with two different microscopes. The photographs were developed from Panatomic (FX-135 mm) film. Pollen grain sizes were observed using 400–600 × magnification; the fine morphological observations of sculpturing and structural features were made at 1000–1200 × magnification (oil immersion).

The numerical aperature (N.A.) of the Zeiss Jena bright field and phase contrast microscope is 1.30 and that of the Reichert interference contrast microscope is 1.25. Interference contrast microscopy is an invaluable aid in discerning details of structure and sculpture, particularly in families such as the Urticaceae and Gramineae where little contrast is available from the specimen to differentiate fine details.

Pollen for the scanning electron microscope (SEM) was prepared by the procedure described by Baum, Bassett, and Crompton (1970). To prevent the collapsing of the thin-walled grains in the Urticaceae, Betulaceae, and Pinaceae, additional steps were taken by fixing in formalin — acetic acid — alcohol (FAA), dehydrating in an ascending ethyl alcohol series, freeze-drying, or critical point drying prior to carbon and gold coating. The photographs were taken on a Mk IIa Steroscan (Cambridge Scientific Instruments Ltd., England) with Polaroid film Type 42.

Voucher slides of all the material examined have been filed in the permanent pollen collection, Biosystematics Research Institute, Canada Department of Agriculture, Ottawa, Ont.

Key to the Pollen Classes*

A. Grains in groups of 4 .... **1. Tetrads**
A. Grains single (monads)
  B. Apertures absent (or indistinct)
    C. With two bladders .... **2. Vesiculate**
    C. Without bladders .... **3. Inaperturate**
  B. Apertures present
    D. Apertures simple, i.e., either pores or colpi
      E. With pores

      1 pore .... **4. Monoporate**
      2 pores .... **5. Diporate**
      3 pores .... **6. Triporate**
      4 pores .... **7. Tetraporate**
      More than 4 pores, equatorially arranged .... **8. Stephanoporate**
      More than 4 pores, some or all not equatorial .... **9. Periporate**

      E. With colpi

      1 colpus .... **10. Monocolpate**
      3 meridional colpi .... **11. Tricolpate**

    D. Apertures compound, i.e., pores in colpi

      3 colpi meridionally arranged .... **12. Tricolporate**
      More than 3 colpi meridionally arranged .... **13. Stephanocolporate**
      More than 3 colpi not meridionally arranged .... **14. Pericolporate**

*Adapted from Faegri and Iversen (1964) and McAndrews et al. (1973).

41

Key to the Spore Classes

A. Without apertures (alete), *Equisetum* .... **3. Inaperturate**

A. With triradiate scar, *Lycopodium* .... **15. Trilete**

Keys to the Pollen and Spores of 145 Canadian Taxa

1. **Tetrads**

    A. All united 4 grains arranged in one plane (tetragonal or rarely rhomboidal) .... ***Typha latifolia***

    A. All united 4 grains arranged in two planes (tetrahedral) .... ***Luzula multiflora***

2. **Vesiculate**

    A. Grains without a constriction between the body and bladders .... ***Picea*** spp.

    A. Grains with a constriction between the body and bladders

        B. Grains including bladders usually more than 90 μm in length

            C. Bladders not folded up underneath the body, the body not forming a saucer or bowl .... ***Abies*** spp.

            C. Bladders usually folded up underneath the body, the body forming a saucer or bowl .... ***Tsuga mertensiana***

        B. Grains including bladders usually less than 90 μm in length .... ***Pinus*** spp.

3. **Inaperturate**

    A. Grains less than 35 μm in diam

        B. Sculpturing basically smooth .... ***Thuja*** spp.

        B. Sculpturing not basically smooth

            C. Grains saucer-shaped or bowl-shaped .... ***Chamaecyparis nootkatensis***

            C. Grains spheroidal, not saucer-shaped or bowl-shaped

                D. Surface with tuberculate bumps in clusters ...... ***Juniperus communis***

                D. Surface without tuberculate bumps in clusters .... ***Taxus brevifolia***

                    E. Sculpturing scabrate .... ***Populus*** spp.

                    E. Sculpturing reticulate .... ***Triglochin maritima***

    A. Grains more than 35 μm in diam

        F. Grains usually 35–45 μm in diam

      G. Irregular ovoid to pear-shaped .... ***Carex aquatilis***

      G. Spheroidal .... ***Equisetum arvense***

   F. Grains usually more than 45 μm in diam

      H. Grains usually 50–80 μm in diam, saucer-shaped or bowl-shaped

         I. Sculpturing rugulate .... ***Tsuga canadensis***
                                                                                                                         T. heterophylla

         I. Sculpturing scabrate to psilate .... ***Larix*** spp.

      H. Grains usually 85–105 μm in diam, not saucer-shaped or bowl-shaped .... ***Pseudotsuga menziesii***

4. Monoporate

   A. Grains 20–45 μm in diam; annulus distinct or indistinct, 2.5–5.0 μm in diam

      B. Grains less than 25 μm in diam, annulus indistinct; sculpturing reticulate .... ***Typha angustifolia***

      B. Grains more than 25 μm in diam, annulus distinct; sculpturing microechinate or microgranulate .... ***Agropyron repens***
                                                                                                        Dactylis glomerata
                                                                                                          Phleum pratense
                                                                                                             Poa pratensis

   A. Grains 80–125 μm in diam; annulus distinct, 6–7 μm in diam .... ***Zea mays***

5. Diporate

   A. Grains less than 19 μm in diam; sculpturing not microechinate under the SEM

      B. Grains 12–13 μm in diam; fine sculpturing evenly distributed over the surface .... ***Laportea canadensis***

      B. Grains 13.5–15.5 μm in diam; coarse sculpturing unevenly distributed over the surface .... ***Pilea pumila***

   A. Grains more than 19 μm in diam; sculpturing microechinate under the SEM .... ***Morus rubra***

6. Triporate

   A. Grains with vestibulate pores .... ***Betula*** spp.

   A. Grains without vestibulate pores

      B. Grains less than 18 μm in diam .... ***Urtica*** spp.
                                                                                                       Parietaria pensylvanica

    B. Grains more than 18 μm in diam

        C. Grains 21–35 μm in diam (polar view)

            D. Ektexine twice as thick as the endexine

                E. Pores aspidate .... ***Cannabis sativa***

                E. Pores not aspidate .... ***Myrica*** spp.

            D. Ektexine and endexine of equal thickness

                F. Sculpturing scabrate .... ***Ostrya virginiana***

                F. Sculpturing scabrate-rugulate .... ***Corylus cornuta***
                                                                                                Celtis occidentalis

        C. Grains 39–52 μm in diam (polar view) .... ***Carya*** spp.

7. **Tetraporate**

    Grains 27–34 μm, av 31 μm in diam, sculpturing scabrate .... ***Comptonia peregrina***

8. **Stephanoporate**

    A. Grains 19–27 μm in diam (polar view); sculpturing scabrate or scabrate-rugulate .... ***Alnus*** spp.

    A. Grains 28.0–33.5 μm in diam (polar view); 5 pored grains with 1 pore not equally distributed on the equator, no arci; sculpturing rugulate to reticulate .... ***Ulmus*** spp.

9. **Periporate**

    A. Grains with less than 30 pores

        B. Grains spheroidal with evenly distributed pores

            C. Sculpturing verrucate .... Plantaginaceae

            C. Sculpturing microechinate

                D. Grains with annulate pores

                    E. Pore number not exceeding 14 .... Plantaginaceae

                    E. Pore number 14–25 .... ***Sarcobatus vermiculatus***
                                                                               (Chenopodiaceae)

                D. Grains with nonannulate pores

                    F. Pores 4–7 μm in diam .... ***Arenaria serpyllifolia***

                    F. Pores 7–10 μm in diam .... ***Thalictrum dasycarpum***

    B. Grains oblate with annulate pores in heteropolar arrangement .... ***Juglans*** **spp.**

 A.   Grains with more than 30 pores

    G. Grains reticulate, 40 μm in diam; pores recessed in the luminae .... ***Polygonum lapathifolium***

    G. Grains microechinate, less than 30 μm in diam .... Chenopodiaceae Amaranthaceae

10. Monocolpate

   Grains often folded showing a furrow .... ***Ginkgo biloba***

11. Tricolpate

 A.   Sculpturing reticulate

    B. Grains oblate in equatorial view, dimensions not exceeding 20 μm; colpi not ending distinctly in the polar area, merging irregularly into the ektexine .... ***Platanus occidentalis***

    B. Grains prolate in equatorial view, dimensions 22–44 μm; colpi ending distinctly in the polar area

       C. Grains 33–38 μm in length .... ***Cardamine pratensis***

       C. Grains 22–32 μm in length

          D. Polar area index 0.8–1.2 .... ***Acer*** spp.
              (excluding ***A. spicatum*** and ***A. circinatum***)

          D. Polar area index 0.18–0.6 .... ***Salix discolor***

 A.   Sculpturing not reticulate

    E. Sculpturing irregular verrucate; grains oblate in equatorial view .... ***Quercus*** spp.

    E. Sculpturing striate or striate-rugulate; grains prolate in equatorial view .... ***Acer*** spp.
          (excluding ***A. spicatum*** and ***A. circinatum***)

12. Tricolporate

 A.   Grains spheroidal; shallow flat colpi .... ***Rumex*** spp.

 A.   Grains not spheroidal; deep conspicuous colpi

    B. Grains triangular or pyramidal .... ***Shepherdia argentea***

    B. Grains not triangular or pyramidal

       C. Grains prolate in equatorial view

    D. Equatorial axis short

        E. Sculpturing pattern distinct, striate or striate-rugulate; grains 20–35 μm in length .... ***Acer circinatum***
                                                                                                                                                       A. spicatum
        E. Sculpturing pattern not distinct; grains 11–16 μm in length .... ***Castanea dentata***

    D. Equatorial axis long .... ***Shepherdia canadensis***

  C. Grains oblate in equatorial view

    F. Sculpturing reticulate with prominent thickened endexine around the pores .... ***Tilia americana***

    F. Sculpturing not reticulate without prominent thickened endexine around the pores

        G. Sculpturing echinate or microechinate .... ***Ambrosia*** spp.
                                                                                                    Iva spp.
                                                                                               Xanthium spp.
                                                                                               Artemisia spp.
                                                                                              Solidago spp.

        G. Sculpturing verrucate or rugulate

            H. Sculpturing verrucate; grains less than 30 μm in diam .... ***Quercus*** spp.

            H. Sculpturing verrucate-rugulate; grains usually more than 30 μm in diam .... ***Fagus*** spp.

13. Stephanocolporate

  A. Grains prolate, the intercolpium area diamond-shaped ....
.... ***Sanguisorba canadensis***

  A. Grains oblate to spheroidal, the intercolpium area mostly pentagonal .... ***Rumex acetosella***
                                                                                                                                                       R. obtusifolia

14. Pericolporate

Grains 28–37 μm, av 33 μm in diam; sculpturing microreticulate ....
.... ***Rumex orbiculatus***

15. Trilete

Spores 23–27 μm, av 25 μm in diam (polar view); sculpturing reticulate on the distal part .... ***Lycopodium selago***

DESCRIPTIONS OF NONFUNGUS SPORES AND POLLEN GRAINS OF VASCULAR PLANTS

The family name is shown in bold type and capital letters at the beginning of each description. The scientific name of the species appears in bold italic type and is followed by the common names in roman type with the preferred name first.

The descriptions are presented in the following order: fern allies, gymnosperms, and angiosperms.

FERN ALLIES

These are not leafy in habit as are the leafy true ferns and they reproduce by spores.

EQUISETACEAE

Equisetum arvense L. Field horsetail, horsetail, devil's-guts, horsepipes, mares tail, snake grass.

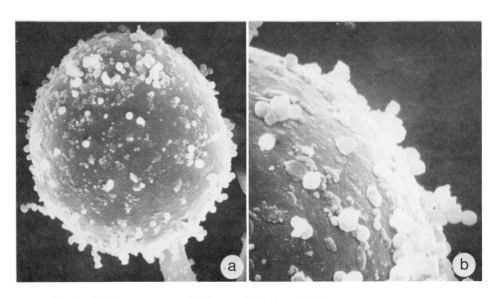

Fig. 6. *Equisetum arvense.* SEM, *a*, × 1640; *b*, × 4200.

Description. Grains inaperturate; spheroidal, 33–44 μm, av 40 um in diam, sculpturing basically smooth except for scattered tuberculate granules, some grains possessing a crumpled outer envelope of clear tissue; ektexine and endexine not separable, the wall 1.5 μm thick; structure intectate; spores not quite mature possessing two hydroscopic spiral bands with flattened ends encircling the grains (elaters).

SEM: Traces of the outer envelope can be seen and the tuberculate granules spread unevenly on an otherwise smooth exine.

Fruiting. Mainly during April and May.

Native to. North America.

Distribution. British Columbia to Newfoundland and the Yukon and Northwest Territories.

Notes. Several spores from *Equisetum* spp. have been caught on exposed slides. There is no information available that the spores from this taxon cause hay fever.

LYCOPODIACEAE

Lycopodium selago L. Mountain club-moss, fir club-moss, running club-moss, coral or staghorn evergreen, buckhorn, wolf's claws.

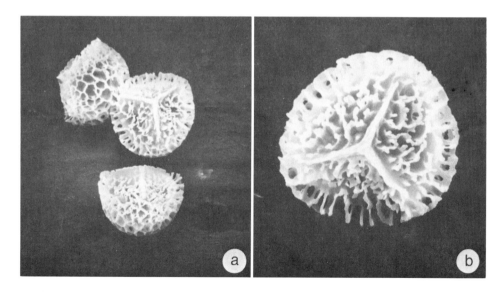

Fig. 7. *Lycopodium selago.* SEM, *a,* × 1200; *b,* × 2400.

Description. Spores trilete, rounded triangular in equatorial contour, in equatorial view 28–33 μm, av 30.5 μm in diam; in polar view 23–27 μm, av 25 μm in diam; sculpturing reticulate on the distal part, reticulum incomplete near the equator and trilete scar; ektexine 3 μm thick, endexine 1 μm thick, complete wall 4 μm thick; structure tectate baculate on the distal part of the grain; structure intectate near the equator and around the proximal part and trilete scar.

SEM: The reticulum is irregular in size and is supported by a row of narrow columellae; the muri are thin and acute, the luminae 3–7 μm in diam; a segment around the equator appears to be definitely intectate possibly due to some cause of development in the spore mother cell.

Fruiting. July 1 to September 15.

Native to. North America.

Distribution. British Columbia to Newfoundland and the Yukon and Northwest Territories.

Notes. Wodehouse (1971) reported that the spores of *L. clavatum,* a close relative of *L. selago,* are available commercially as lycopodium powder. The powder is highly flammable and has been used in fireworks and stage lightning. It is also used in theatrical makeup and in pharmacy. Wodehouse mentioned a case of a pharmacist who developed severe rhinorrhea and asthma from using lycopodium powder for dusting pills. Recently Hyde (1972) mentioned that the spores of club-moss, *L. clavatum,* are allergenic. Experiments by Linskens and Jorde (1974) with spores of *L. clavatum* confirmed the existence in man of a persorption process and a paracellular uptake of spore particles via the digestive tract and their importance in the induction of hay fever.

GYMNOSPERMS

These are shrubs and trees, which have needlelike or scalelike leaves and reproduce by seeds. All the families, genera, and species are arranged in alphabetic order.

CUPRESSACEAE

Chamaecyparis nootkatensis (D. Don) Spach   Yellow cypress, Alaska cedar.

Description. Grains inaperturate, spheroidal; mostly saucer-shaped, bowl-shaped, or folded with a ruptured tear in the exine, 25–34 μm, av 29 μm in diam; sculpturing with irregularly spaced tuberculate particles on the surface; wall 1.0–1.5 μm thick; structure intectate.

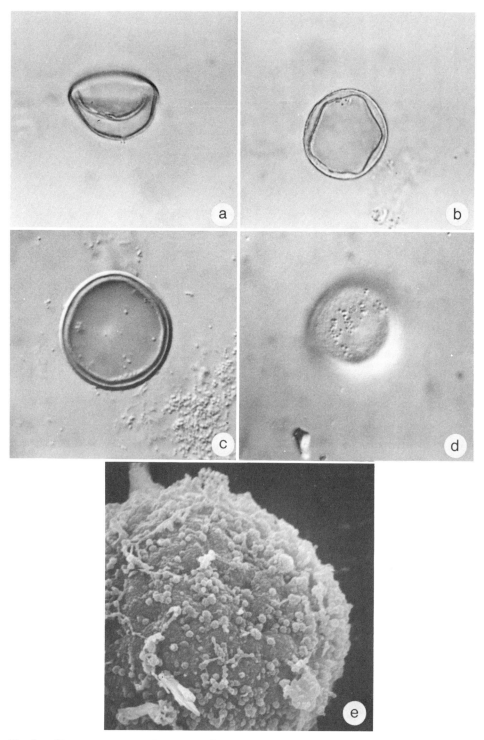

Fig. 8. *Chamaecyparis nootkatensis. a–b*, light microscope, × 1000; *c–d*, interference contrast, × 1000; *e*, SEM, × 3990.

SEM: Sculpturing very similar to the pollen of *Taxus canadensis*.

Flowering. April with the peak period occurring around the middle of the month (Bassett and Crompton 1966).

Native to. Western North America.

Distribution. Mostly along the western coastline of British Columbia.

Notes. It is likely that the pollen causes hay fever when shed in large amounts. Wodehouse (1971) mentioned that the pollen of closely related species causes hay fever.

CUPRESSACEAE

Juniperus communis L. Common juniper, spreading juniper, juniper.

Description. Grains inaperturate, spheroidal; occasionally split showing a fissura as in *Pseudotsuga menziesii,* 21–28 µm, av 25 µm in diam; sculpturing on the surface smooth with occasional tuberculate bumps in clusters; wall 1.5 µm thick; structure intectate.

SEM: Sculpturing evenly distributed spheroidal microtubercles, 0.6 µm in diam, on a smooth exine surface.

Flowering. The latter part of March, and in April and May, depending on the elevation and latitude.

Native to. North America.

Distribution. Every province and the Northwest Territories.

Notes. The pollen of *Juniperus virginiana* L., eastern red cedar; *J. scopulorum* Sarg., western juniper; and *J. horizontalis* Moench, creeping juniper, are very similar to the grains of the common juniper under the light microscope. The pollen of *J. scopulorum* under the SEM differs from the pollen of *J. communis* by having unevenly distributed spheroidal microtubercles less than 0.4 µm in diam that possess microechinate spines.

Pollen of the junipers in large amounts may cause hay fever.

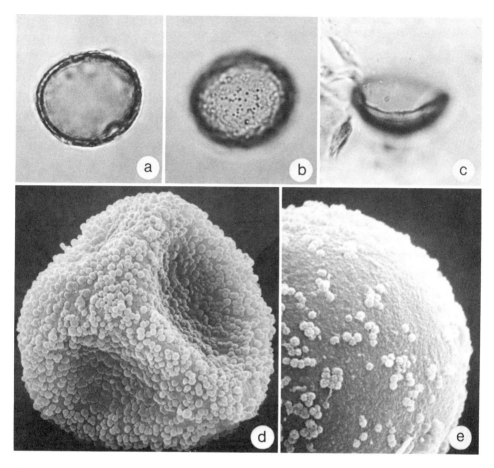

Fig. 9. *Juniperus* spp. *a–c, J. communis*, light microscope, × 1000; *d, J. communis*, SEM, × 3675; *e, J. scopulorum*, SEM, × 4060.

CUPRESSACEAE

Thuja occidentalis L.   Eastern white cedar, white arbor-vitae, cedar, eastern cedar.

Description.   Grains spheroidal to irregular in shape (especially if the grains are acetolyzed), 25–34 μm, av 29 μm in diam; sculpturing smooth on the surface with irregular spaced clusters of microtubercles (microtubercles fewer in number than on *Juniperus* grains); wall 0.5 μm thick; structure intectate.

SEM: Surface of the grains similar to that of *T. plicata*.

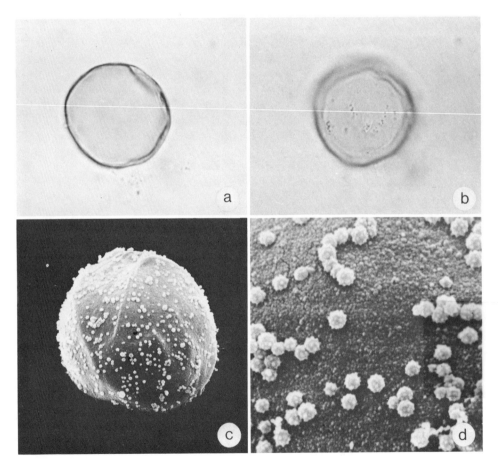

Fig. 10.   *Thuja* spp. *a–b*, light microscope, × 1000; *c*, SEM, × 1750; *d*, SEM, × 8400.

Flowering. May.

Native to. Eastern North America.

Distribution. Southwestern Manitoba to Nova Scotia.

Notes. The pollen on *T. plicata* Donn, western red cedar, is very similar to that of the eastern white cedar. When their pollen is produced in large amounts it may cause hay fever (see notes under *Thuja plicata*).

CUPRESSACEAE

Thuja plicata Donn   Western red cedar, giant arbor-vitae.

Description. Grains similar to those of *Thuja occidentalis.*

SEM: The exine surface is composed of very fine granular processes; more or less echinate tubercles approximately 0.7 μm in diam are unevenly distributed on top of the granular surface.

Flowering. April and May.

Native to. Western North America.

Distribution. Coastal British Columbia and the mountain ranges in the interior.

Notes. Conifers with few exceptions (mountain cedar, eastern red cedar, and incense cedar) have not been considered to be significant allergy producers in this country. Recent studies have indicated that the pollen extracts of a large number of species of the cypress family are immunogenic and cross reactive in man and rabbits. This suggests their possible allergenic importance to our mobile population in many parts of Canada and the world (Tai-June et al. 1974).

GINKGOACEAE

Ginkgo biloba L.   Ginkgo, maidenhair tree.

Description. Grains monocolpate; oblate or boat-shaped, 15–16 μm wide × 27–45 μm long, av 23 × 36 μm; grains often folded showing a furrow extending the total length of the grain; sculpturing somewhat undulating, not distinctive; ektexine twice as thick as the endexine, complete wall 1.0–1.5 μm thick; structure tectate.

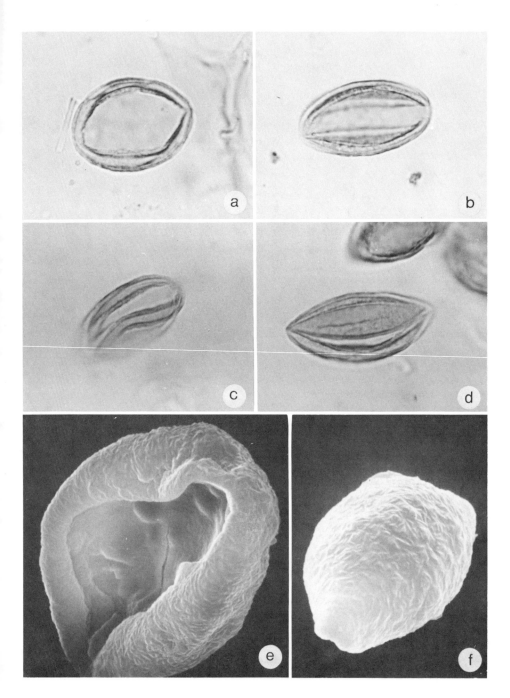

Fig. 11. *Ginkgo biloba. a–d*, light microscope, × 1000; *e–f*, SEM, × 3500.

SEM: Grains show a rugulate sculpturing on the distal side away from the furrow with an undulating exine.

Flowering. May.

Native to. China.

Distribution. Occasionally cultivated in British Columbia, Ontario, Quebec, and the Maritime Provinces.

Notes. The pollen is not known to cause hay fever.

Pollen Key to the Species of *Abies* (Firs)

A. Grains mostly with a triradiate mark at the center of the dorsal cap

    B. Dorsal cap smooth, diam less than 5 μm .... **Abies balsamea**

    B. Dorsal cap with an undulating surface; diam more than 5 μm
    .... **A. grandis**

A. Grains without a triradiate mark at the center of the dorsal cap
.... **A. lasiocarpa**

PINACEAE

Abies balsamea (L.) Mill.   Balsam fir, fir.

Description.   Grains vesiculate, disaccate, usually with a constriction between the body and bladders; colpi clearly observed in ventral and meridional views; grains 81–115 µm, av 94 µm long (including bladders); main body of grain 66–91 µm, av 78 µm long (excluding bladders); some grains exhibiting a slight triradiate scar on the dorsal cap; dorsal cap 2.5–4.5 µm thick; sculpturing on bladder surface finely reticulate; exine structure on the dorsal cap smooth, rodlike pilae.

   SEM: See *Abies grandis*, Fig. 12.

Flowering.   May and early June.

Native to.   North America.

Distribution.   Alberta to Newfoundland, mainly in the boreal area.

Notes.   Although the pollen of *Abies* spp. is shed in large amounts, it is not known to cause hay fever. Adams and Morton (1972) have an excellent SEM photograph of balsam fir showing the rounded body and bladders.

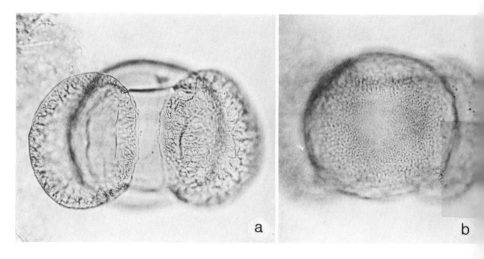

Fig. 12.   *Abies* spp. *a–b*, *A. balsamea*, light microscope, × 400. *c–d*, *A. lasiocarpa*, SEM *c*, × 1680, *d*, × 590. *e–g*, *A. grandis*, note thin scar area on cap; *e–f*, light microscope × 400; *g*, SEM, × 735.

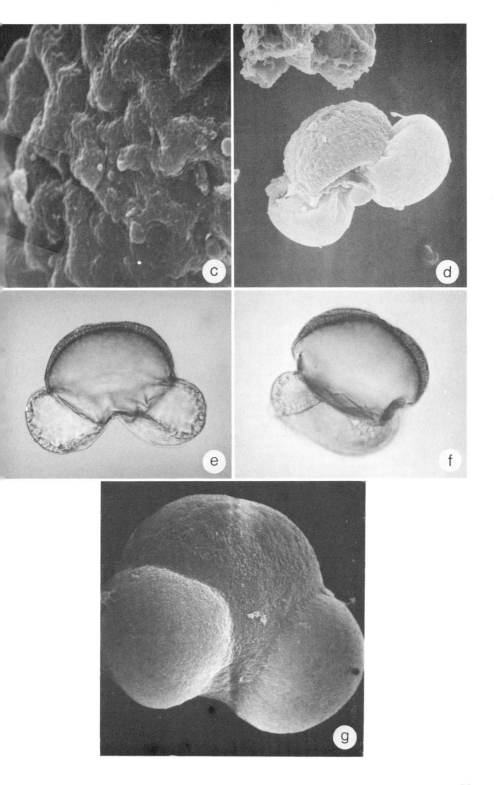

PINACEAE

Abies grandis (Dougl.) Lindl.   Grand fir, lowland fir.

Description.   Grains vesiculate, disaccate, usually with a constriction between the body and bladders, colpi clearly observed in ventral and meridional views; grains 100–130 μm, av 119 μm long (including bladders); main body of grain 82–104 μm, av 90 μm long (excluding bladders); many grains exhibiting a prominent triradiate scar at the apex of the dorsal cap in the form of a long three-parted furrow extending to the edge of the slight marginal ridge; dorsal cap 5–7 μm thick; sculpturing on bladder surface similar to *A. balsamea;* exine structure, rodlike pilae.

   SEM: Sculpturing on dorsal cap, showing strong rugulate markings.

Flowering.   April and the early part of May.

Native to.   Western North America.

Distribution.   Coastline and the interior of southern British Columbia.

Notes.   Although large amounts of pollen grains are shed from this species, it is not known to cause hay fever.

PINACEAE

Abies lasiocarpa (Hook.) Nutt.   Alpine fir, subalpine fir, Rocky Mountain fir.

Description.   Grains vesiculate, disaccate, usually with a constriction between the body and bladders, colpi clearly observed in ventral and meridional views; grains 110–130 μm, av 120 μm long (including bladders); main body of grain 80–115 μm, av 95 μm long (excluding bladders); no triradiate scar visible on dorsal cap; dorsal cap 4–5 μm thick; sculpturing on bladder surface similar to *A. balsamea;* exine structure, uneven rodlike pilae.

   SEM: Sculpturing on dorsal cap showing small evenly spaced pila and visible remnant of scar.

Flowering.   May into July depending on the elevation.

Native to.   Western North America.

Distribution.   Northwest Territories, Rocky Mountains of British Columbia and Alberta.

Notes. Several botanists consider *A. lasiocarpa* to be a subspecies of *A. balsamea*, but the pollen morphology of both species suggests that they are distinct from each other. There is no information that the pollen of this species causes hay fever.

Pollen Key to the Species of *Larix* (Larches)

A. Grains about 80 μm wide, 89 μm long .... **Larix lyallii**

A. Grains 60–65 μm wide; 68–73 μm long

    B. Wall 1.5–2.0 μm thick .... **L. laricina**

    B. Wall 2.0–2.5 μm thick .... **L. occidentalis**

PINACEAE

Larix laricina (Du Roi) K. Koch Tamarack, eastern larch, Alaska larch, hackmatack.

Description. Grains inaperturate, spheroidal to oblate, often saucer-shaped, bowl-shaped, or folded as in *Ginkgo* spp. with a ruptured exine; triradiate scar evident in most grains; grains 51–75 μm, av 62 μm wide, 56–80 μm, av 68 μm long; wall 1.5–2.0 μm thick; sculpturing showing very minute scabrate processes in optical section; structure intectate.

    SEM: Sculpturing composed of irregularly shaped and densely packed micropilae that are visible at only high magnification.

Flowering. May.

Native to. North America.

Distribution. All provinces and the Northwest Territories.

Notes. In the southern part of its range this tree is rather common in bogs and marshy areas. There is no information that the pollen of *Larix* spp. causes hay fever.

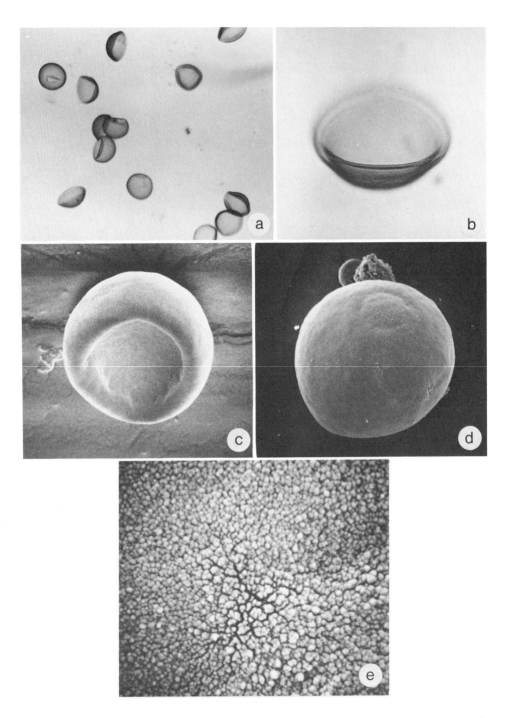

Fig. 13. *Larix* spp. *a–b*, light microscope, *a*, × 200; *b*, × 500; *c–e*, SEM, *c*, × 800, *d*, × 800, *e*, × 8900.

PINACEAE

Larix lyallii Parl.   Alpine larch, subalpine larch, Lyall's larch, tamarack.

Description.   Grains inaperturate, spheroidal to oblate, often saucer-shaped, bowl-shaped, or folded as in *Ginkgo* spp. with a ruptured exine; triradiate scar evident on all grains; grains 74–89 µm, av 80 µm wide; 84–96 µm, av 89 µm long; sculpturing not clearly defined; wall 2.0–2.5 µm thick; structure undetermined.

SEM: Sculpturing similar to the pollen of *L. laricina. See* Fig. 13.

Flowering.   April to June, depending on climate as influenced by elevation factors.

Native to.   Western North America.

Distribution.   Subalpine forests of the Rocky Mountains in southern British Columbia and Alberta.

Notes.   The pollen is the largest of the three larch species examined.

PINACEAE

Larix occidentalis Nutt.   Western larch, western tamarack, tamarack.

Description.   Grains inaperturate; spheroidal to oblate, bowl-shaped or folded as in *Ginkgo* spp. with a ruptured exine; triradiate scar evident on all grains; grains 50–76 µm, av 65 µm wide; 65–84 µm, av 73 µm long; wall 2.0–2.5 µm thick; sculpturing scabrate, undulating; structure same as in grains of *L. laricina.*

SEM: Sculpturing similar to the pollen of *L. laricina. See* Fig. 13.

Flowering.   Latter part of April and into May.

Native to.   Western North America.

Distribution.   Southeastern British Columbia.

Notes.   The triradiate scar is more prominent in this taxon than in the pollen of *L. laricina.*

Pollen Key to the Species of *Picea* (Spruces)

A. Grains with a shallow sinus area formed by colpi and nonpendulous bladders, sculpturing unevenly arranged and granular on dorsal cap

   B. Grains more than 100 μm in length (including bladders) .... ***Picea glauca***

   B. Grains less than 100 μm in length (including bladders)

      C. Grains with microverrucate particles in the sinus area, dorsal cap 4.5–6.0 μm thick .... ***P. rubens***

      C. Grains with prominent verrucate particles in the sinus area, dorsal cap 4.0 μm or less thick .... ***P. mariana***

A. Grains with a deep sinus area formed by colpi and pendulous bladders, sculpturing of regular, matlike micropilae on dorsal cap.... ***P. rubens* × *P. mariana***

PINACEAE

Picea glauca (Moench) Voss   White spruce, cat spruce.

Description. Grains vesiculate, disaccate, without a constriction between the body and bladders, colpi clearly observed in ventral and meridional views in the bladder sinus (space between the bladders and the colpi); grains 85–129 μm, av 100 μm long (including bladders); dorsal cap 4 μm thick; no bladder attachments or a distinctive marginal crest; sculpturing granular on dorsal cap; sculpturing indistinct on bladder surface; bladders fine inner supporting weblike reticulum.

    SEM: Sculpturing on the dorsal cap, densely packed, irregular clusters of microverrucate particles.

Flowering. The latter part of May and into June.

Native to. North America.

Distribution. All provinces and the Northwest Territories, common in the Boreal Forest Region.

Notes. Pollen from the spruces is not known to cause hay fever.

    Hybridization occurs between the following species: *P. glauca* × *P. sitchensis*; *P. glauca* × *P. engelmanii*; *P. sitchensis* × *P. engelmanii*; and *P. rubens* × *P. mariana*. Hybrids are often more dominant than the parental species. Pollen grains from these hybrids are difficult to identify.

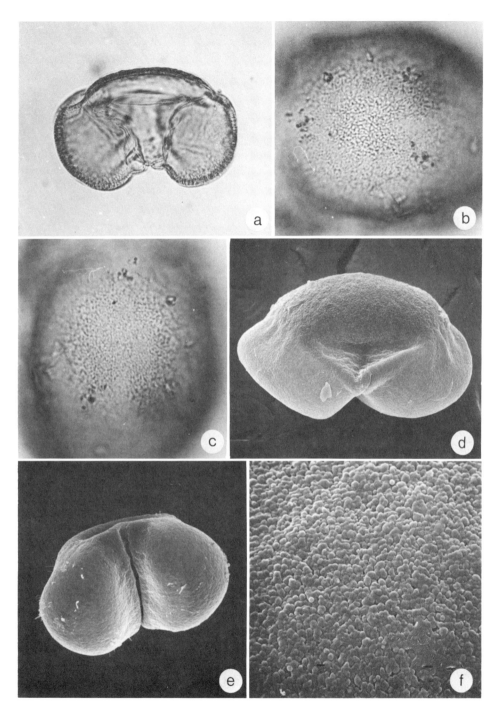

Fig. 14. *Picea glauca. a–c*, light microscope, *a*, × 400, *b–c*, high and low focus of dorsal cap, × 1000; *d–f*, SEM, *d–e*, × 750, *f*, dorsal cap, × 3800.

PINACEAE

Picea mariana (Mill.) B.S.P.   Black spruce, bog spruce, swamp spruce.

Description. Grains vesiculate, disaccate, without a constriction between the body and bladders, colpi clearly observed in ventral and meridional views; sinus shallow between bladders and colpi with strongly marked verrucate grains 79–89 µm, av 84 µm long (including bladders); particles up to the colpi's edge; dorsal cap 4.0 µm thick; sculpturing small irregular granules on the dorsal cap; sculpturing indistinct on bladder surface; bladders a fine inner supporting weblike reticulum.

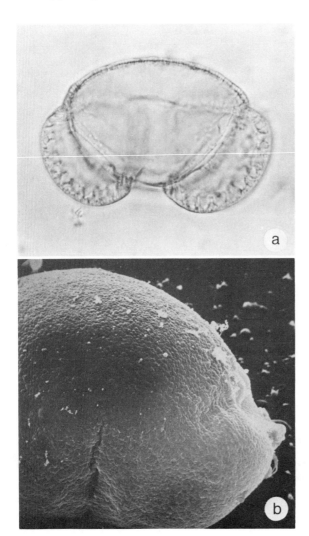

Fig. 15. *Picea mariana. a*, light microscope, × 1000; *b*, SEM, × 1500.

SEM: Sculpturing similar to that of *P. glauca*.

Flowering. Latter part of May and in June.

Native to. North America.

Distribution. All provinces and northward to the Northwest Territories.

Notes. See notes under *P. glauca*.

The hybrid *P. mariana* × *P. rubens* occurs mainly in eastern Ontario, southern Quebec, and the Maritime Provinces. 50–70% of the pollen grains produced by artificial hybrids from the parental stock are 83–94 μm, av 89 μm long (including bladders). There are deeply sinuate spaces between pendulous drooping bladders; sculpturing on the dorsal cap is composed of matlike pilae (Fig. 17). Sculpturing, as seen under the SEM, on the dorsal cap is composed of a matlike arrangement of various sized pilae.

Feinberg (1946) noted that the asthma condition of two patients became greatly aggravated around Christmas time, and their attacks were proved to be due to the drying needles of the Christmas tree, *P. mariana*.

Dr. David Malloch (Wyse and Malloch 1970) reported that a surprising number of persons suffer from Christmas tree allergy. The most likely causes are the tree resins that give Christmas trees their smell. Several people developed symptoms either while decorating the tree or within 24 h. Others developed symptoms 3 to 4 days after the tree was set up. The Christmas trees investigated were also found to release weed, grass, and tree pollen grains into the house. It is thought that the pollen grains had collected on the branches during the summer and were released as the tree dried out. Studies of the possible introduction of mold into house air by Christmas trees were inconclusive.

PINACEAE

Picea rubens Sarg.   Red spruce, yellow spruce.

Description.   Grains vesiculate, disaccate, without a constriction between the bladders; colpi clearly observed in ventral and meridional views; sinus shallow between bladders and colpi with slightly verrucate particles near and within the colpi; grains 86–97 μm, av 93 μm long (including bladders); dorsal cap 4.5–6.0 μm thick; grains without bladder attachments or a distinctive marginal crest; sculpturing evenly distributed microgranules on dorsal cap; sculpturing indistinct on bladder surface; bladders have an inner weblike reticulum.

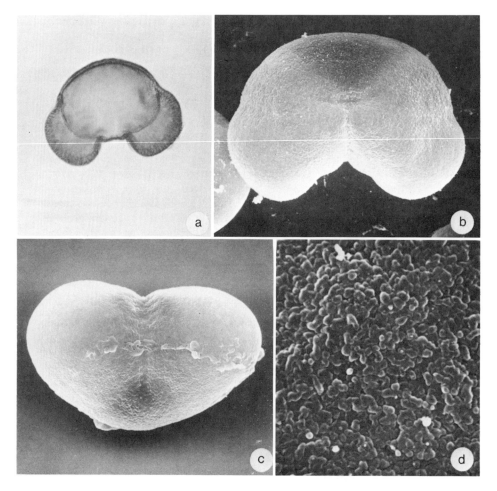

Fig. 16.   *Picea rubens. a*, light microscope, × 400; *b–d*, SEM, *b*, × 900, *c*, × 840, *d*, × 3850.

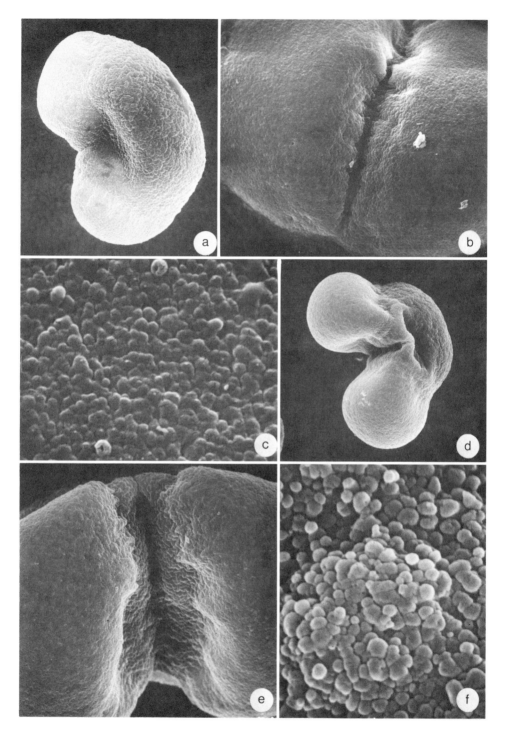

Fig. 17. *Picea marina*, and *P. marina* × *P. rubens*. *a–f*, SEM; *a–c*, *P. marina*, *a*, × 840, *b*, × 1580, *c*, × 7350; *d–f*, *P. mariana* × *P. rubens*, *d*, × 840, *e*, × 1750, *f*, × 8250.

SEM: Sculpturing similar to that of *P. glauca.*

Flowering. May and early June.

Native to. Eastern North America.

Distribution. Ontario to Nova Scotia.

Notes. See notes under *P. glauca.*

Pollen Key to the Species of *Pinus* (Pine)

A. Grains diploxylon
    B. Grains with distinct bladder attachments .... **Pinus resinosa**
    B. Grains without distinct bladder attachments
        C. Grains (including bladders) mostly over 75 μm long
            D. Dorsal cap less than 2 μm thick .... **P. rigida**
            D. Dorsal cap more than 2 μm thick .... **P. contorta** var. **latifolia**
                                                                                                                    P. ponderosa
        C. Grains (including bladders) mostly less than 75 μm long .... **P. contorta** var. **contorta**

A. Grains haploxylon
    E. Grains with prominent bladder attachments
        F. Grains (including bladders) mostly more than 75 μm long .... **P. flexilis**
                                                                                                     P. monticola
        F. Grains (including bladders) mostly less than 75 μm long .... **P. strobus**
    E. Grains with inconspicuous bladder attachments
        G. Dorsal cap more than 2.5 μm thick .... **P. banksiana**
        G. Dorsal cap less than 2.5 μm thick .... **P. albicaulis**

PINACEAE

Pinus albicaulis Engelm. Whitebark pine, scrub pine, white stemmed pine.

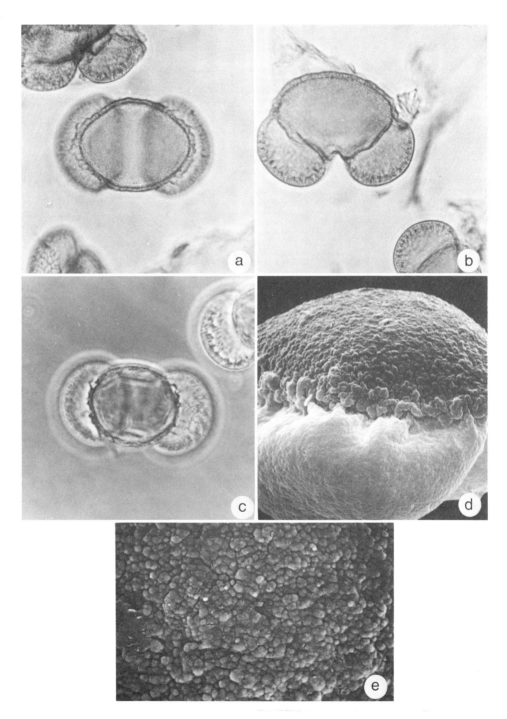

Fig. 18. *Pinus albicaulis. a–b*, light microscope, × 400; *c*, phase contrast, × 400; *d–e*, SEM, *d*, bladders, attachments and dorsal cap, × 1680, *e*, middle of dorsal cap, × 4200.

Description. Grains vesiculate, disaccate, and haployxlon, usually with a constriction between the body and bladders; grains 59–85 μm, av 73 μm long (including bladders), main body 45–66 μm, av 55 μm long (excluding bladders); dorsal cap 2 μm thick; bladder attachments inconspicuous, marginal ridge somewhat toothed; sculpturing fine granular on dorsal cap, sculpturing indistinct on bladder surface.

SEM: Dorsal cap, fine irregular verrucate particles; marginal crest at the bladders of verrucate clusters.

Flowering. Mostly in July.

Native to. Western North America.

Distribution. British Columbia and Alberta in the Subalpine Forest Region.

Notes. Whitebark pine overlaps with limber pine in distribution. The pollen is easily separated from limber pine by morphological characters. See notes under *P. banksiana* on pine pollen causing hay fever.

PINACEAE

Pinus banksiana Lamb.   Jack pine, Banksian pine, scrub pine.

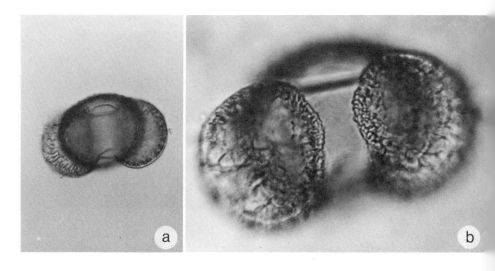

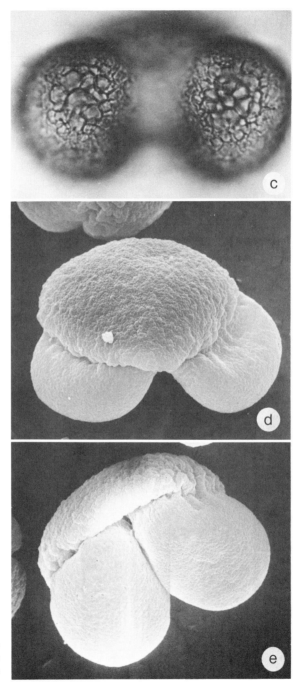

Fig. 19. *Pinus banksiana.* a–c, light microscope, *a*, × 400, *b*, × 1000, *c*, × 1000; *d–e*, SEM, × 1575.

Description. Grains vesiculate, disaccate, and haploxylon, usually with a constriction between the body and bladders; grains 46–64 µm, av 58 µm long (including bladders); main body 39–51 µm, av 43 µm long (excluding bladders); dorsal cap 3 µm thick; bladder attachments inconspicuous, marginal ridge slightly noticeable; sculpturing fine granular on dorsal cap, sculpturing indistinct on bladder surface.

SEM: Dorsal cap, granular groupings of microverrucate particles.

Flowering. Latter part of May and June with a peak period the first 2 weeks of June.

Native to. North America.

Distribution. The Mackenzie District, Northwest Territories, to Nova Scotia, exceptionally abundant northwest of Lake Superior.

Notes. Hybrids between *Pinus banksiana* and *P. contorta* Dougl. var. *latifolia* Engelm., lodgepole pine, are known from Western Canada where the two taxa overlap in their distribution. No attempt has been made to examine the pollen grains of the hybrids.

There have been a few reports that the pollen of pine species causes hay fever (Wodehouse 1971) and asthma (Newmark and Itkin 1967).

PINACEAE

Pinus contorta Dougl. var. *latifolia* Engelm.   Lodgepole pine, western jack pine, cypress.

Description. Grains vesiculate, disaccate, and diploxylon, usually with a constriction between the body and bladders; grains 72–85 µm, av 77 µm long (including bladders); main body 48–57 µm, av 56 µm long (excluding bladders); dorsal cap 3 µm thick; bladder attachments inconspicuous, marginal rim enlarged at the dorsal root of bladder and main body; sculpturing fine granular clusters on surface of dorsal cap.

SEM: Sculpturing undulating clusters of microverrucate particles, roughened marginal rim on dorsal cap; sculpturing indistinct on bladder surface.

Flowering. May and June.

Native to. Western North America.

Distribution. The interior of southern British Columbia to southwestern Saskatchewan, Yukon, and Northwest Territories.

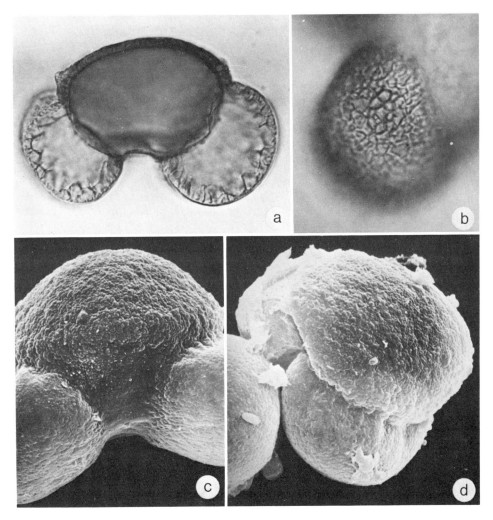

Fig. 20. *Pinus contorta* var. *latifolia*. a–b, light microscope, a, × 900, b, × 1000; c–d, SEM, c, × 1750, d, × 1675.

Notes. The pollen of *Pinus contorta* Dougl. var. *contorta*, shore pine, is very similar to the pollen of lodgepole pine except for the size of the grain. Pollen grains of *P. contorta* are 51–72 μm, av 61 μm long (including bladders); main body 41–54 μm, av 47 μm long (excluding bladders). The range of *P. contorta* is along the coastline of British Columbia to Alaska. See notes under *P. banksiana* on pine pollen causing hay fever.

PINACEAE

Pinus flexilis James   Limber pine, Rocky Mountain white pine.

Description. Grains vesiculate, disaccate, and haploxylon, usually with a constriction between the body and bladders; grains 73–102 μm, av 88 μm long (including bladders), main body 49–63 μm, av 57 μm long (excluding bladders); dorsal cap 2.5–3.5 μm thick; bladders attached to grain by buttresslike ridges; sculpturing granular on dorsal cap, sculpturing indistinct on bladder surface.

SEM: The cap is composed of verrucate-rugulate ridges that are irregularly arranged; these ridges become more prominent at the marginal crest near the bladders. Grains viewed in a polar dorsal position show large bladder attachments.

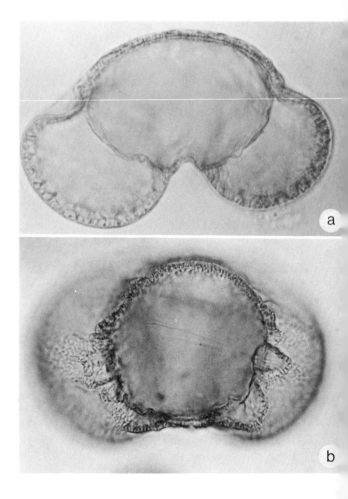

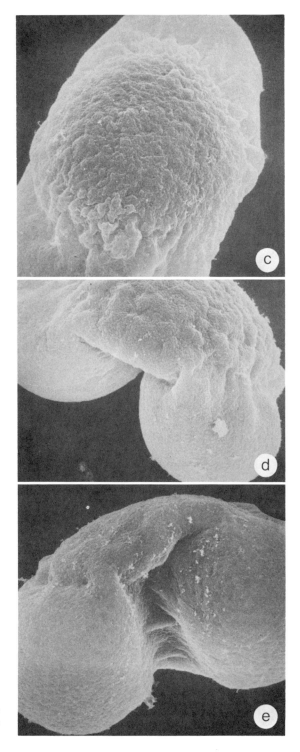

Fig. 21. *Pinus flexilis.* a–b, light microscope, a, × 900, b, × 900; c–e, SEM, × 1400.

Flowering. Late June with the peak period following shortly afterwards.

Native to. Western North America.

Distribution. Western and eastern foothills of the Rocky Mountains in British Columbia and Alberta.

Notes. Because the pollen of this species is the largest of all the pine pollen examined, it is the easiest to separate from the other taxa. See notes under *P. banksiana* on pine pollen causing hay fever.

PINACEAE

Pinus monticola Dougl.   Western white pine, silver pine.

Description. Grains vesiculate, disaccate, and haploxylon, usually with a constriction between the body and bladders; grains 78–96 μm, av 86 μm long (including bladders); main body 49–65 μm, av 58 μm long (excluding bladders); dorsal cap 3 μm thick; bladders attached to grain by buttresslike thickenings at the dorsal root and marginal crest; sculpturing granular uneven particles on dorsal cap, sculpturing smooth on bladder surface.

   SEM: Grains with prominent bladder attachments; dorsal cap, strongly verrucate undulating elements; colpi bordered by verrucate particles.

Flowering. End of April and in May.

Native to. Western North America.

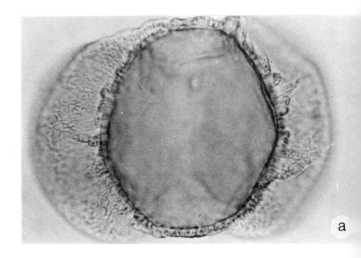

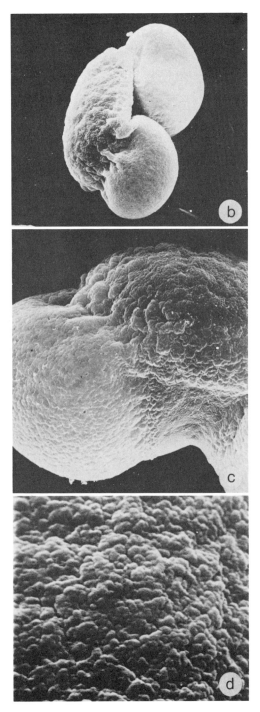

Fig. 22. *Pinus monticola.* a, light microscope, × 900; b–d, SEM, b, × 770, c, × 1680, d, × 4400.

Distribution. The southwestern coastline of British Columbia to the Rocky Mountains in Alberta.

Notes. Pollen grains are very similar to those of *P. flexilis.* See notes under *P. banksiana* on pine pollen causing hay fever.

PINACEAE

Pinus ponderosa Laws.   Ponderosa pine, yellow pine, ball pine.

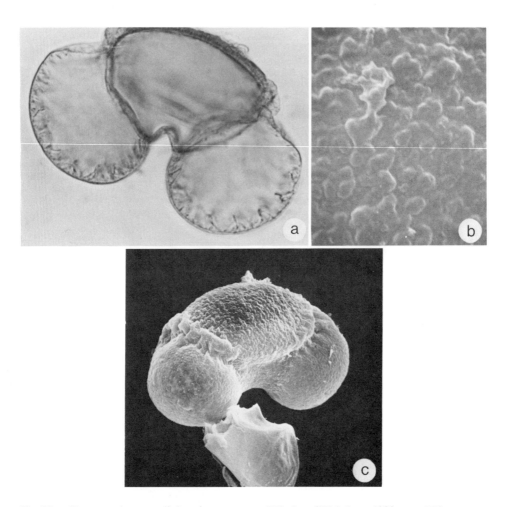

Fig. 23.   *Pinus ponderosa. a,* light microscope, × 900; *b–c,* SEM, *b,* × 4300, *c,* × 750.

Description. Grains vesiculate, disaccate, and diploxylon, usually with a constriction between the body and bladders; grains 80–98 μm, av 85 μm long (including bladders); main body 53–69 μm, av 57 μm long (excluding bladders); dorsal cap 2.5 μm thick; bladder attachments not conspicuous; marginal crest at dorsal root thick and prominent; sculpturing verrucate particles on dorsal cap, sculpturing indistinct on bladder surface.

SEM: Sculpturing of irregular verrucate particles separated by valla on dorsal cap; sculpturing undulating wrinkled on the bladder surface; marginal ridge prominent and thickened above the dorsal root of the bladders.

Flowering. May and June.

Native to. Western North America.

Distribution. Dry regions of British Columbia.

Notes. The sculpturing on the dorsal cap of the grains is the most distinctive of all the Canadian pines. See notes under *P. banksiana* on pine pollen causing hay fever.

PINACEAE

Pinus resinosa Ait.   Red pine, Norway pine, yellow pine, Canadian red pine.

Description. Grains vesiculate, disaccate, and diploxylon, usually with a constriction between the body and bladders; grains 53–73 μm, av 63 μm long (including bladders); main body 38–58 μm, av 48 μm long (excluding bladders); dorsal cap 2.5–3.5 μm thick; exine in the germinal area between the bladders sometimes destroyed after being acetolyzed; bladder attached to grain by buttresslike bridges merging with the marginal ridge at the dorsal root of the bladder; sculpturing coarse granular on dorsal cap, sculpturing fine granular on bladder surface.

SEM: Dorsal cap sculpturing, granular groupings of verrucate particles, no indication of a germinal furrow; front edge of dorsal cap smooth; surface rough and distinct on bladder attachments.

Flowering. Latter part of May and June, with a peak period generally the first 2 weeks of June.

Native to. North America.

Distribution. Southern Manitoba to Newfoundland, most common in Great Lakes–St. Lawrence Forest Region on sandy, gravelly soils.

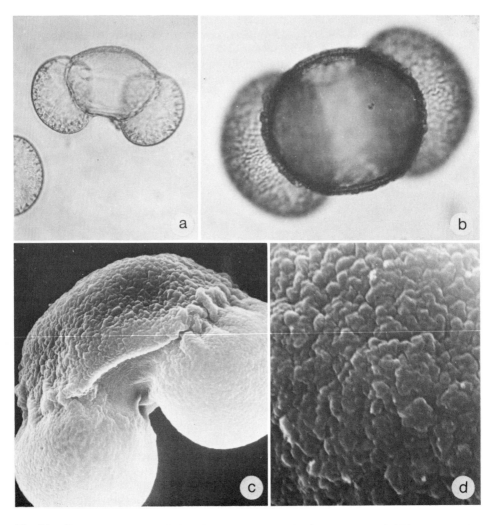

Fig. 24. *Pinus resinosa.* a–b, light microscope, a, × 400, b, × 1000; c–d, SEM, c, × 1680, d, 4200.

Notes. Pollen from *Pinus* spp. is a source of honey dew yielding black honey (Pellett 1947). See notes under *P. banksiana* on pine pollen causing hay fever.

PINACEAE

Pinus rigida Mill.   Pitch pine, rigid pine, scrub pine.

Description. Grains vesiculate, disaccate, and diploxylon, usually with a constriction between the body and bladders; grains 71–84 μm, av 78 μm long (including bladders); main body 45–59 μm, av 52 μm long (excluding bladders); dorsal cap 15 μm thick; exine in the germinal area between the bladders sometimes missing after being acetolyzed; bladder attachments not prominent and apparently weak because many are detached from the body by acetolysis; marginal ridge not conspicuous; sculpturing clusters of microverrucate particles on dorsal cap, sculpturing indistinct on bladder surface.

SEM: Dorsal cap, small tightly compressed verrucate particles, colpi smooth, bladder attachments not conspicuous.

Flowering. April and May.

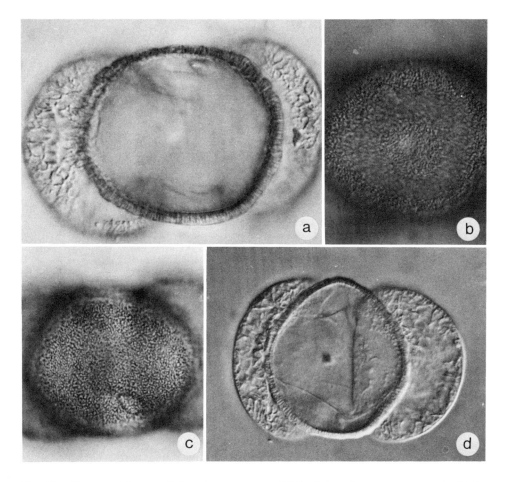

Fig. 25. *Pinus rigida.* a–c, light microscope, × 1000; d, interference contrast, × 1000; e–g, SEM, e, × 385, f, × 830, g, × 4200.

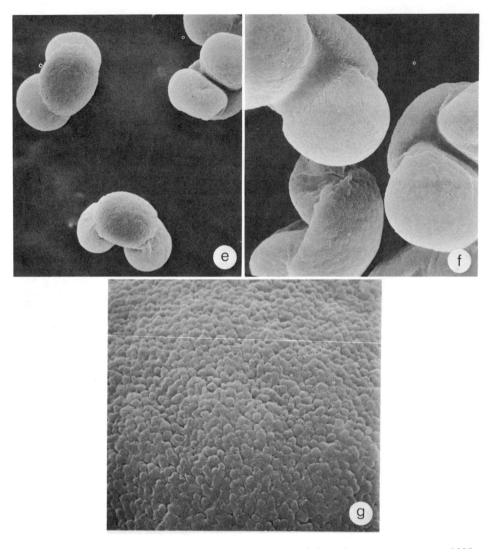

Fig. 25. *Pinus rigida.* a–c, light microscope, × 1000; d, interference contrast, × 1000; e–g, SEM, e, × 385, f, × 830, g, × 4200.

Native to.   Eastern North America.

Distribution.   Southeastern Ontario near the St. Lawrence River and southwestern Quebec.

Notes.   The thickness of the dorsal cap, which is less than 2 μm, separates this taxon from the pollen of the other species. See notes under *P. banksiana* on pine pollen causing hay fever.

PINACEAE

Pinus strobus L.   Eastern white pine, Weymouth pine, white pine.

Description. Grains vesiculate, disaccate, and haploxylon, usually with a constriction between the body and bladders; grains 44–69 µm, av 59 µm long (including bladders); main body 45–52 µm, av 49 µm long (excluding bladders); dorsal cap 2.5–3.0 µm thick; bladders in meridional view attached to the grains by prominent ridges; in optical section view, grains with a distinct marginal ridge at the dorsal root of the bladder; sculpturing fine granular on the body surface, sculpturing indistinct on the bladder surface.

SEM: Dorsal cap, granular groupings of microverrucate particles.

Flowering. Latter part of May and June with a peak period generally the first 2 weeks of June.

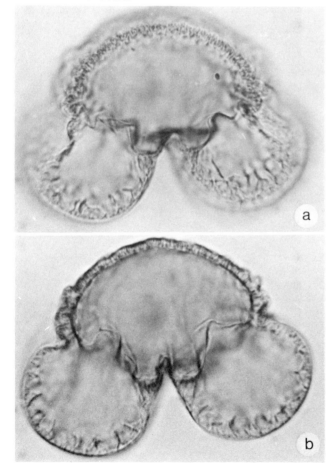

Fig. 26. *Pinus strobus. a–c,* light microscope, × 1000; *d–e,* SEM, *d,* × 1750, *e,* × 4375.

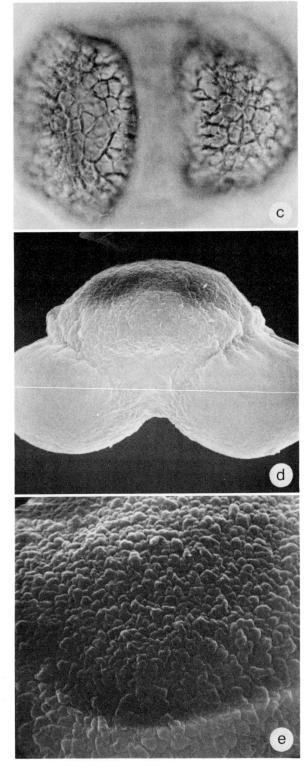

Fig. 26. *Pinus strobus.* a–c, light microscope, × 1000; d–e, SEM, d, × 1750, e, × 4375.

Native to. North America.

Distribution. Southern Manitoba to Newfoundland, common in Great Lakes–St. Lawrence Forest Region.

Notes. The pollen is similar to that of other pine taxa and occasionally causes hay fever.

PINACEAE

Pseudotsuga menziesii (Mirb.) Franco   Douglas-fir, British Columbia fir, red fir, yellow fir, Douglas spruce.

Description. Grains inaperturate, spheroidal, banded equatorially with a thickened ring or ridge forming the base for a supporting triradiate scar over the thinner polar end, surface often rupturing to create a fissura on the thinner end at right angles across the triradiate bands; outer wall of pollen absent after acetolysis; grains 85–105 μm, av 95 μm in diam; sculpturing microtuberculate or in most cases indistinct; wall 2.5 μm at thickened end and 1.5 μm at thinner end; structure intectate.

SEM: Sculpturing very similar to the pollen of *Larix laricina*.

Flowering. Late March and early April.

Native to. Western North America.

Distribution. British Columbia and southwestern Alberta.

Notes. No information is available as to whether or not the pollen of this tree causes hay fever.

(Fig. 27 overleaf)

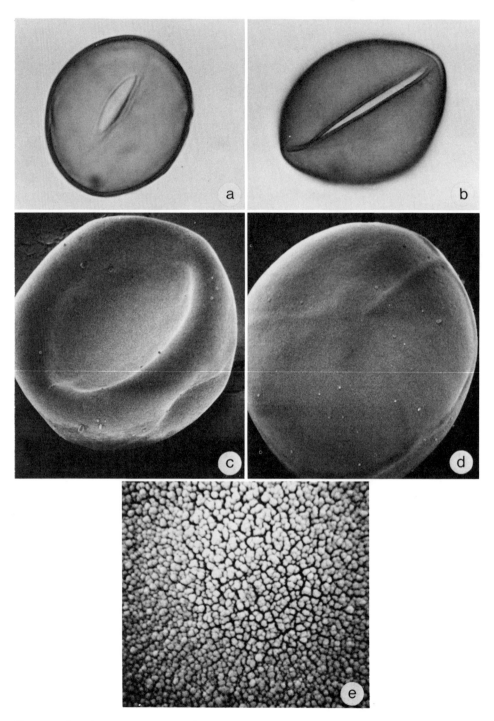

Fig. 27. *Pseudotsuga menziesii. a–b*, light microscope, × 400; *c–e*, SEM, *c–d*, × 800, *e*, × 8400.

Pollen Key to the Species of *Tsuga* (Hemlocks)

A. Grains inaperturate, saucer-shaped or bowl-shaped, without bladders

    B. Sculpturing rugulate with distinctive microechinate processes .... ***Tsuga heterophylla***

    B. Sculpturing rugulate with no distinctive microechinate processes .. ***T. canadensis***

A. Grains vesiculate .... ***T. mertensiana***

PINACEAE

Tsuga canadensis (L.) Carr.   Eastern hemlock, hemlock spruce.

Description.   Grains inaperturate, circular in polar view, heteropolar; vesiculate with two rudimentary bladders (bladders not always conspicuous); grains 62–77 μm long; sculpturing strongly rugulate with the reduced bladders more rugulate than the dorsal cap; wall 4.5–6.5 μm thick; structure intectate.

    SEM: Sculpturing rugulate, no microechinate processes on the muri on either the dorsal or ventral surface; grains bowl-shaped with a coarse rugulate marginal rim.

Flowering.   May.

Native to.   Eastern North America.

Distribution.   Western Ontario to Newfoundland, common in southern parts.

Notes.   In spite of their large size and weight, pollen grains are frequently caught on atmospheric pollen slides. According to Wodehouse (1971), they have occasionally been suspected of causing hay fever.

(Fig. 28 overleaf)

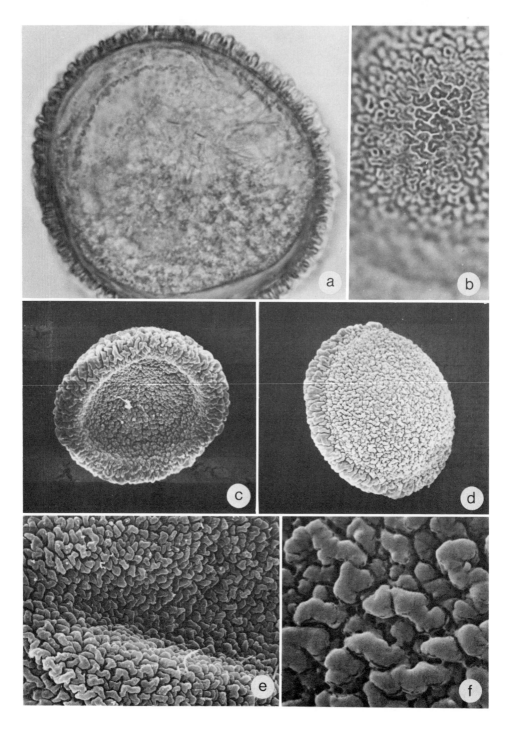

Fig. 28. *Tsuga canadensis. a–b*, light microscope, × 1000; *c–f*, SEM, *c–d*, × 1100, *e*, × 1600, *f*, × 7350.

PINACEAE

Tsuga heterophylla (Raf.) Sarg. Western hemlock, British Columbia hemlock, Alaska pine.

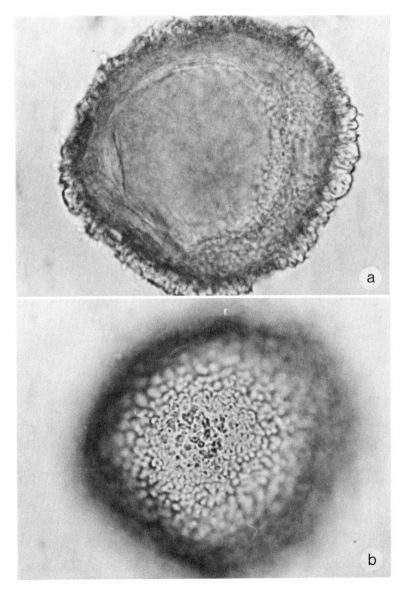

Fig. 29. *Tsuga heterophylla.* a–b, light microscope, × 1000; c–e, SEM, c, × 770, d, × 7280, e, × 1575.

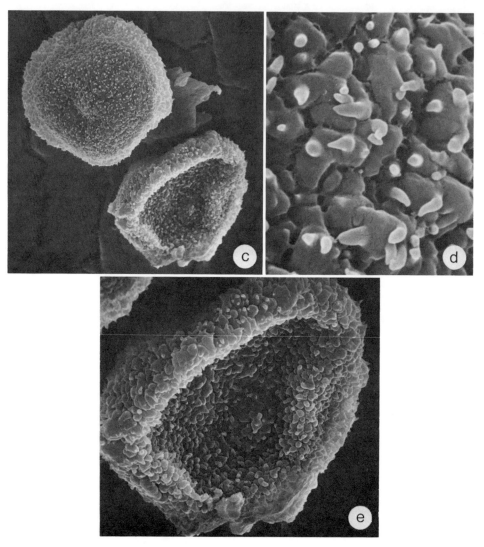

Fig. 29. *Tsuga heterophylla.* a–b, light microscope, × 1000; c–e, SEM, c, × 770, d, × 7280, e, × 1575.

Description. Grains inaperturate, often saucer-shaped, bowl-shaped, or folded, circular in polar view; grains 61–79 µm, av 69 µm long; sculpturing rugulate on dorsal cap, topped with microspines; sculpturing rugulate on ventral surface, microspines less evident.

SEM: Sculpturing rugulate with distinctive microechinate processes on muri mostly on dorsal surface; grains bowl-shaped with a coarse rugulate marginal rim.

Flowering. April and May.

Native to. Western North America.

Distribution. British Columbia and Yukon, common along the Pacific coastline.

Notes. The pollen is easily separated from the pollen of the other two species by the distinctive microspines on the surface. When the pollen of this species is produced in large amounts, it may cause hay fever.

PINACEAE

Tsuga mertensiana (Bong.) Carr.   Mountain hemlock, black hemlock.

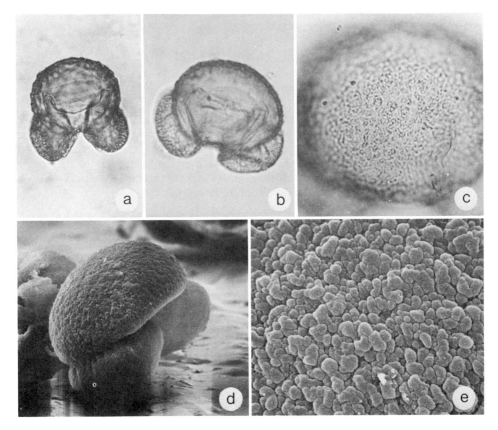

Fig. 30. *Tsuga mertensiana.* a–c, light microscope, a–b, × 400, c, × 1000; d–e, SEM, d, × 770, e, × 3675.

Description. Grains vesiculate, disaccate, saucer-shaped or bowl-shaped, bladders folded underneath the body; grains 90–110 μm, av 100 μm long; sculpturing slightly undulating, microverrucate on dorsal cap; wall of dorsal cap 2.0–2.5 μm thick; structure intectate.

SEM: Sculpturing densely packed, verrucate, microglomerules; bladders with no distinctive marginal rim.

Flowering. April and May with the peak period usually around the middle of May.

Native to. Western North America.

Distribution. British Columbia and the Yukon, mainly in the coastal regions.

Notes. The pollen has distinctive bladders, which separate it from the pollen of the other two species.

When the pollen is produced in large amounts, it may cause hay fever.

TAXACEAE

Taxus brevifolia Nutt.   Western yew, Pacific yew.

Description. Grains irregular, rarely spheroidal in shape, inaperturate; exine often ruptured by acetolysis; grains 22–29 μm, av 25 μm in diam; unacetolyzed grains with an irregular mass (contraction of the exine) at the center, acetolyzed grains often with a triradiate, double folded pinch somewhat like the scar on some fern spores; sculpturing microverrucate with irregular spaced elements; wall about 1 μm thick; structure intectate.

SEM: Sculpturing similar to the pollen of *Juniperus communis.*

Flowering. Latter part of April and May.

Native to. North America.

Distribution. British Columbia and Alberta, mostly along the western coastline and in the Rocky Mountains.

Notes. It is practically impossible to separate the pollen from that of *T. canadensis* Marsh., ground hemlock, under the light microscope, but under the SEM the pollen is similar to that of *Juniperus scopulorum* Sarg. and *J. communis* L. Pollen of *Taxus* spp. was caught in very small amounts on slides exposed in 1965 at 17 stations in British Columbia (Bassett and Crompton 1966).

It is doubtful if the pollen of yews causes hay fever.

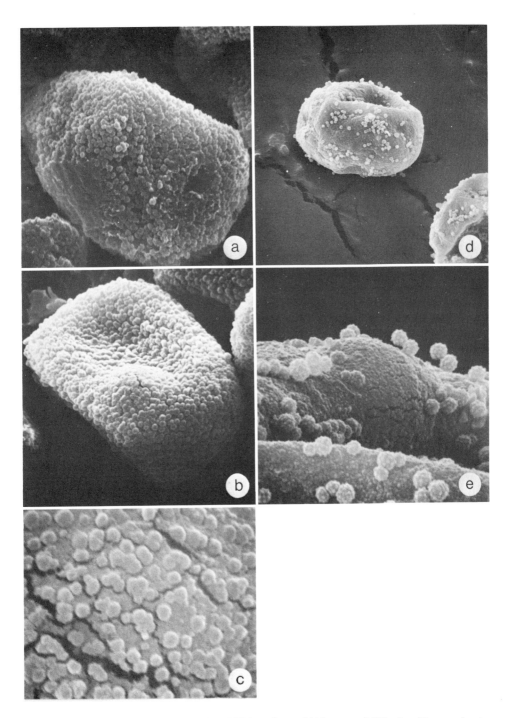

Fig. 31. *Taxus* spp. *a–c, T. brevifolia,* SEM, *a–b,* × 3400, *c,* × 8400. *d–e, T. canadensis,* SEM, *d,* × 1600, *e,* × 9290.

ANGIOSPERMS

These are herbs, shrubs, and trees, which have many forms of leaves and reproduce by seeds. All the families, genera, and species are arranged in alphabetic order with the exception of the family Compositae. In this family the order is *Ambrosia* spp., *Iva* spp., *Solidago canadensis,* and *Xanthium* spp. (ragweeds and their relatives) followed by *Artemisia* spp. (the sages or wormwoods).

Pollen Key to the Species of *Acer* (Maples)

A. Grains tricolporate, with striate sculpturing

    B. Ektexine and endexine distinct and of equal thickness; grains 20.5–27.0 µm × 22.0–35.0 µm ....*Acer circinatum*

    B. Ektexine and endexine indistinct; grains 14.0–19.0 µm × 18.0–24.0 µm ....*A. spicatum*

A. Grains tricolpate, with striate or rugulate sculpturing

    C. Sculpturing striate

        D. Polar equatorial index 1.0–1.2

            E. Grains 36–44 µm long ....*A. macrophyllum*

            E. Grains 24–30 µm long

                F. Surface with compound striations under SEM ....*A. glabrum* var. *douglasii*

                F. Surface with simple striations under SEM ....*A. rubrum*

        D. Polar equatorial index less than 1.0 ....*A. saccharum*

        G. Sculpturing rugulate under the SEM ....*A. nigrum*

        G. Sculpturing striate-rugulate or foveolate-rugulate under the SEM

            H. Sculpturing striate-rugulate ....*A. pensylvanicum**

            H. Sculpturing foveolate-rugulate ....*A. negundo*
A. saccharinum

**A. pensylvanicum* occasionally has equatorial constrictions or marks in the colpi resembling pores. In some of the grains the sculpturing pattern is not clear.

ACERACEAE

Acer circinatum Pursh   Vine maple, mountain maple.

Description.   Grains tricolporate, polar equatorial index 1.2; prolate, rarely spheroidal in equatorial view, 20.5–27.0 μm × 22.0–35.5 μm; sculpturing striate; ektexine and endexine of equal thickness, wall about 2.5 μm thick; structure semitectate.

SEM: Colpi about 25 μm long, pores about 2.5 μm long; sculpturing striate with many elongated, wormlike, divided, simple ridges; scattered perforated openings in the luminae between the ridges.

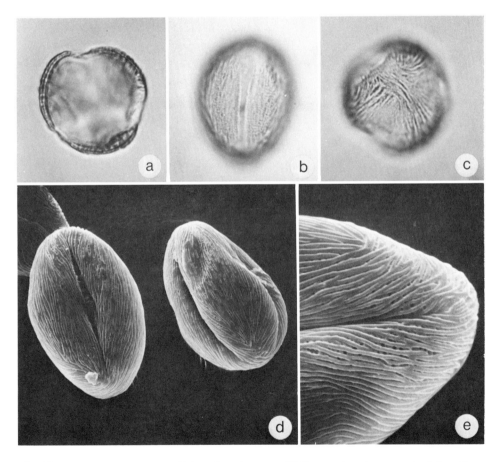

Fig. 32.   *Acer circinatum.* *a–c*, light microscope, × 1000, *a*, polar view, equatorial section, *b*, equatorial view of colpi, *c*, surface of polar end; *d–e*, SEM, *d*, × 1400, *e*, × 3500.

Flowering. End of April to the end of May, with peak about the first week of May.

Native to. Western North America.

Distribution. Southwestern British Columbia, a constituent of the forest understory (Hosie 1969).

Notes. Because vine maple is rather sporadic in its distribution and mostly insect-pollinated, its pollen is not an important factor in causing hay fever.

ACERACEAE

Acer glabrum Torr. var. *douglasii* (Hook.) Dipp. Douglas maple, Rocky Mountain maple.

Description. Grains tricolpate, polar equatorial index 1.0; prolate, rarely spheroidal in equatorial view, 21.5–26.5 μm × 24.0–31.0 μm; sculpturing coarse striate; structure semitectate.

SEM: Colpi about 25 μm long; sculpturing very coarse striate; ridges compounded of several twisted and fused elements; perforated openings in the luminae.

Flowering. April 15 to May 15 with a peak between the last week of April and the first week of May.

Native to. Western North America.

Distribution. British Columbia and southwestern Alberta, fairly common on Vancouver Island and in the interior of British Columbia.

Notes. Because this small tree or shrub occurs at high elevations and is mostly insect-pollinated, it is doubtful if it is important in causing hay fever.

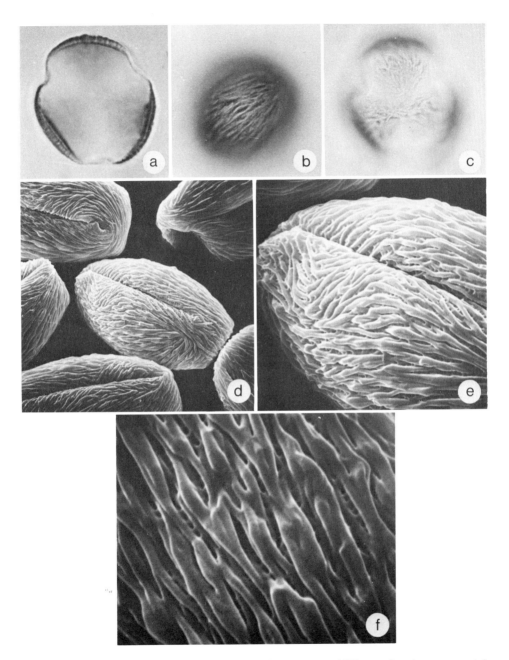

Fig. 33. *Acer glabrum* var. *douglassi*. a–c, light microscope, × 1000, a, polar view, equatorial section, b, equatorial surface, c, surface of polar end; d–f, SEM, d, × 1400, e, × 3500, f, × 7000.

ACERACEAE

Acer macrophyllum Pursh   Bigleaf maple, broadleaf maple, Oregon maple, British Columbia maple.

Description.   Grains tricolpate, polar equatorial index 1.2; prolate in equatorial view, 30.0–41.0 µm × 36.0–44.0 µm; sculpturing coarse striate; ektexine and endexine about 2 µm thick; structure semitectate.

SEM: Colpi about 30 µm long; sculpturing coarse striate; ridges compounded of several, twisted and fused elements; elongated perforated openings in the luminae.

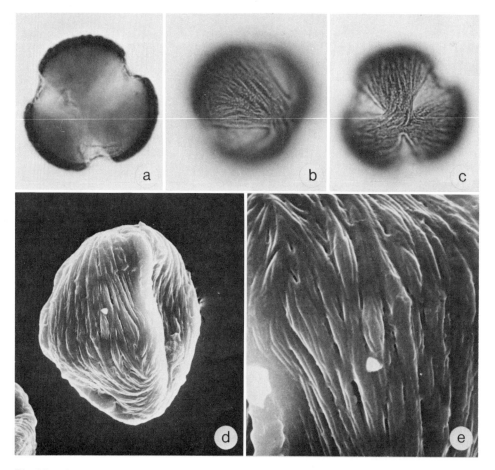

Fig. 34.   *Acer macrophyllum.* a–c, light microscope, × 1000, a, polar view, equatorial section, b, surface striations, c, surface of polar end; d–e, SEM, d, × 1400, e, × 3500.

Flowering. Early April to middle of May with the peak period in the latter part of April and early May.

Native to. Western North America.

Distribution. Southwestern British Columbia, fairly common near the coastline and on Vancouver Island.

Notes. Although this species is mostly insect-pollinated, some pollen becomes airborne and could cause hay fever.

ACERACEAE

Acer negundo L.   Manitoba maple, box-elder, ash-leaved maple, island box-elder, inland Manitoba maple.

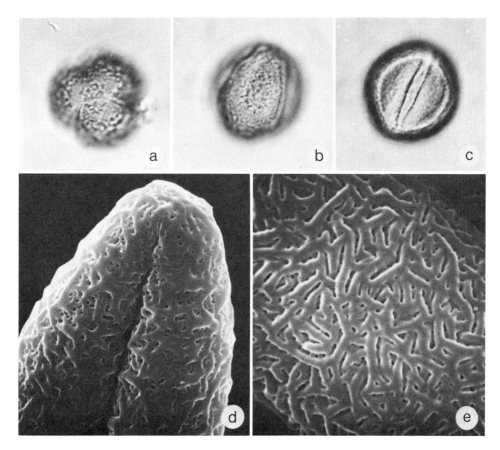

Fig. 35. *Acer negundo.* a–c, light microscope, × 1000; a, polar surface, b, equatorial surface, c, colpi; d–e, SEM, d, × 3500, e, × 7000.

Description. Grains tricolpate, rarely dicolpate, polar equatorial index 1.1; prolate, rarely spheroidal in equatorial view, 24.5–30.0 μm × 27.5–33.5 μm; sculpturing rugulate; ektexine and endexine about 1.5 μm thick; structure semitectate.

SEM: Colpi about 2.6 μm long, surface rough, sculpturing foveolate-rugulate especially near the furrows' edge.

Flowering. At Ottawa, the first mean flowering date May 5, the peak flowering period a few days later; in Western Canada 2 to 3 weeks later.

Native to. North America.

Distribution. Western Alberta to Ontario, common in Saskatchewan and Manitoba, established far beyond its natural range in many parts of Canada.

Notes. This species is wind-pollinated and when in full flower it causes considerable hay fever, especially in Western Canada.

ACERACEAE

Acer nigrum Michx. f.   Black maple, hard maple, rock maple.

Description. Grains tricolpate, polar equatorial index 1.1; prolate, rarely spheroidal in equatorial view, 26.5–32.5 μm × 31.0–37.0 μm; sculpturing reticulate-rugulate; ektexine and endexine about 2 μm thick; structure semitectate.

SEM: Colpi about 25 μm long; sculpturing foveolate-rugulate.

Flowering. Middle of April to the middle of May, maximum flowering period latter part of April and early May.

Native to. North America.

Distribution. Southern Ontario and a small area around Montreal, Que.

Notes. It is similar to sugar maple in that it is pollinated by insects and wind. In areas where this species is common, the airborne pollen may cause hay fever.

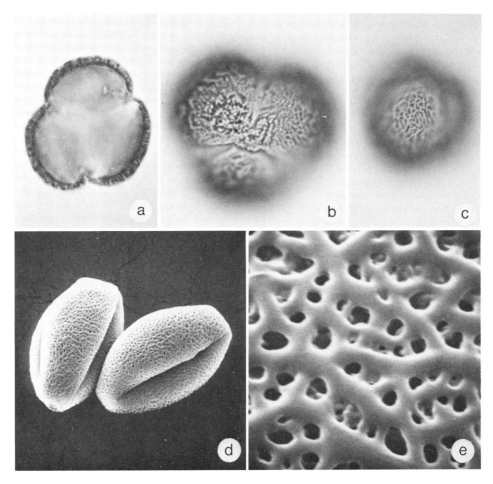

Fig. 36. *Acer nigrum.* a–c, light microscope, × 1000, a, polar view, equatorial section, b, polar surface, c, equatorial surface; d–e, SEM, d, × 1800, e, × 9100.

ACERACEAE

Acer pensylvanicum L.   Striped maple, moosewood, moose maple.

Description.   Grains tricolpate, polar equatorial index 1.0; prolate, occasionally spheroidal in equatorial view, 21.5–25.0 µm × 21.5–26.0 µm; sculpturing striate and occasionally rugulate; ektexine and endexine about 2 µm thick; structure indistinct.

SEM: Colpi about 24 µm long, sculpturing striate and occasionally rugulate with no definite pattern, perforated openings in the luminae.

Flowering.   May 15 to June 15 with maximum flowering period between the last week of May and the first week of June.

Native to.   North America.

Distribution.   Southern Ontario to Nova Scotia, common within the Great Lakes–St. Lawrence and Acadia forest regions.

Notes.   Helmich (1963) stated in his morphological studies of North American maples that the pollen of striped maple is occasionally tricolporate and these studies confirm this. Because striped maple grows throughout the forests of Eastern Canada and is generally a small tree and insect-pollinated, it is doubtful if much pollen travels far from the source. The species does not play an important role in causing hay fever.

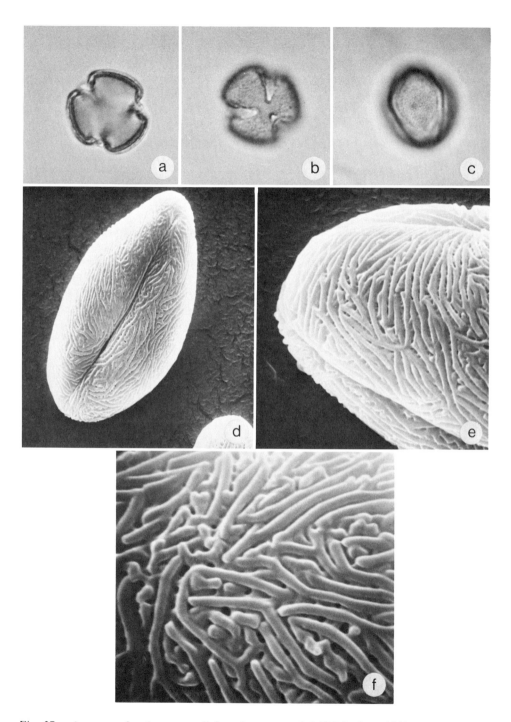

Fig. 37. *Acer pensylvanicum*. a–c, light microscope; d–f, SEM, d, × 1800, e, × 4500, f, × 9100.

ACERACEAE

Acer rubrum L.   Red maple, soft maple, scarlet maple, water maple.

Description. Grains tricolpate, polar equatorial index 1.2; prolate, rarely spheroidal in equatorial view, 24.0–30.0 μm × 26.5–30.0 μm; sculpturing striate; ektexine and endexine about 1.5 μm thick; structure semitectate.

SEM: Colpi nearly 30 μm long; sculpturing striate with many wormlike, simple ridges; scattered perforated openings in the luminae.

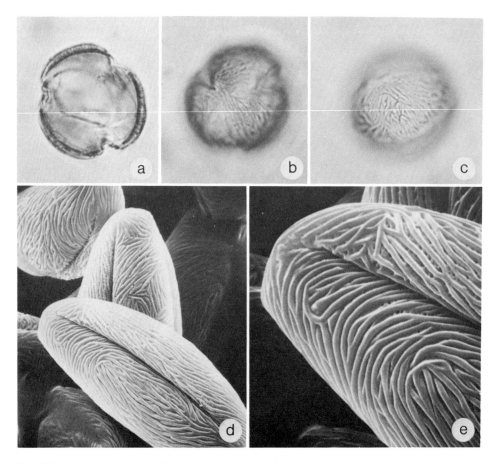

Fig. 38. *Acer rubrum. a–c*, light microscope, × 1000, *a*, polar view, equatorial section, *b*, polar surface, *c*, equatorial surface; *d–e*, SEM, *d*, × 1550, *e*, × 3670.

Flowering. At Ottawa the mean first flowering date April 26 with peak flowering period shortly afterwards, extreme southern part of Ontario the flowering period 1 to 2 weeks earlier.

Native to. North America.

Distribution. Ontario to Newfoundland.

Notes. Where abundant, airborne pollen from this species causes hay fever.

ACERACEAE

Acer saccharinum L.   Silver maple, soft maple, white maple, swamp maple, river maple, water maple.

Description. Grains tricolpate, polar equatorial index 1.0; prolate, rarely spheroidal in equatorial view, 23.5–31.5 µm × 27.5–36.0 µm; sculpturing rugulate; ektexine and endexine about 1.5 µm thick; structure semitectate.

SEM: Colpi about 30 µm long; sculpturing foveolate-rugulate (small pits), especially near the furrows' edge.

Flowering. At Ottawa the mean first flowering date April 11, with the peak flowering period shortly afterwards; in New Brunswick, flowering about 2 weeks later.

Native to. North America.

Distribution. Western border of Ontario to New Brunswick, common in southern Ontario and Quebec.

Notes. It is well known that people suffer from hay fever caused by the pollen of this species. When this tree begins flowering at Ottawa, many enquiries are received concerning hay fever symptoms. The species is pollinated by insects and wind.

(Fig. 39 overleaf)

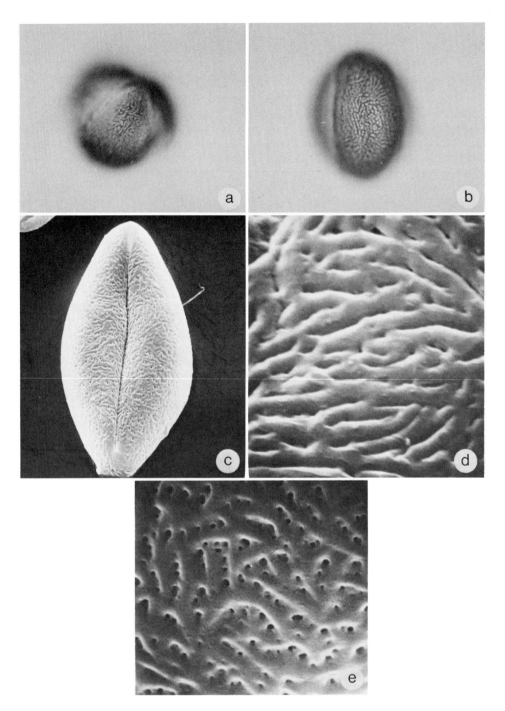

Fig. 39. *Acer saccharinum.* a–b, light microscope, × 1000, a, polar end and colpi, b, equatorial surface; c–e, SEM, c, × 1800, d, polar end, × 9100, e, equatorial surface, × 9100.

ACERACEAE

Acer saccharum Marsh.   Sugar maple, hard maple, rock maple, black maple, curly maple, bird's-eye maple.

Description. Grains tricolpate, polar equatorial index 0.8; prolate, rarely spheroidal in equatorial view, 23.0–28.5 μm × 28.5–35.0 μm; sculpturing striate; ektexine and endexine about 2 μm thick; structure semitectate.

SEM: Colpi about 19 μm long, exine pattern irregular striate swirls, perforated openings in the luminae.

Flowering. The latter part of April and through May; at Ottawa the mean first flowering date May 9 with the peak flowering period following shortly afterwards (Bassett, Holmes, and MacKay 1961).

Native to. North America.

Distribution. Ontario to Nova Scotia (Hosie 1969), most common in southern Ontario and Quebec and some areas of the Maritime Provinces.

Notes. The sugar maple is pollinated by insects and wind. In airborne pollen surveys in Ontario and Quebec many pollen grains from maples have been caught in the spring. A person allergic to maple pollen would probably have symptoms over a long period in the spring because of the successive flowering dates of the various species (Bassett 1956).

(Fig. 40 overleaf)

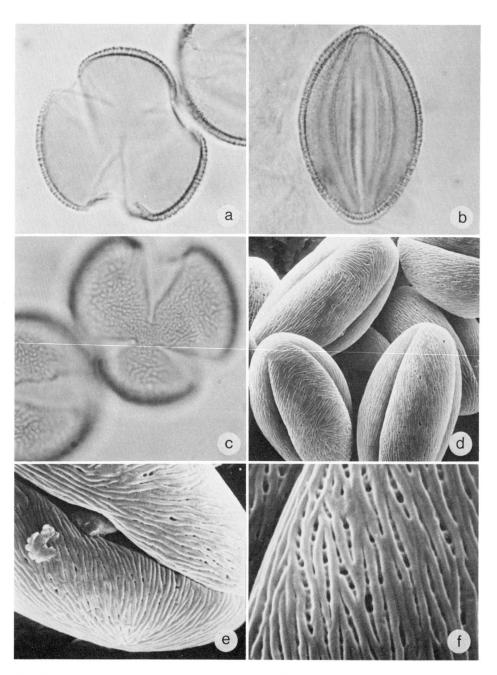

Fig. 40. *Acer saccharum.* a–c, light microscope, × 1000, a, polar view, equatorial section, b, equatorial section, c, polar surface; d–f, SEM, d, × 1800, e, × 4550, f, × 9100.

ACERACEAE

Acer spicatum Lam.   Mountain maple, white maple, dwarf maple.

Description.   Grains tricolporate, polar equatorial index 1.1; prolate, rarely spheroidal in equatorial view, 14.5–19.0 μm × 18.0–24.0 μm; sculpturing striate, ektexine and endexine about 0.5 μm thick; structure semitectate.

SEM: Colpi about 17 μm long, pores about 2 μm long; sculpturing striate; perforated openings in the luminae.

Flowering.   From about last week of May until last week of June, with the maximum flowering during the early part of June.

Native to.   North America.

Distribution.   Western Saskatchewan to Newfoundland, fairly common in woodlands.

Notes.   Because this small tree grows in forest areas, very little of its pollen is carried long distances. Therefore, it is doubtful if it causes much hay fever.

(Fig. 41 overleaf)

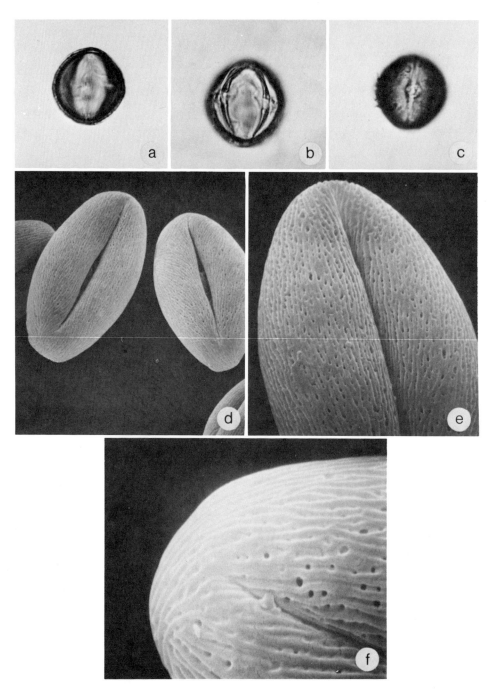

Fig. 41. *Acer spicatum*. a–c, light microscope, × 1000, a, equatorial section, b, colpi margins and pores, c, pore; d–f, SEM, d, × 1750, e, × 4270, f, × 8500.

AMARANTHACEAE

Amaranthus albus L.   Tumble pigweed, tumbleweed.

Description. Grains periporate, spheroidal, 16–23 μm, av 20 μm in diam; pores 30–35, 2 μm in diam, circular in outline, covered with a granular membrane; sculpturing smooth surface flecked with regular spaced microgranules; ektexine and endexine 1 μm thick; structure tectate.

SEM: Sculpturing microechinate, spines evenly distributed over smooth exine surface; pores with an inconsistent number of granular particles on the membrane.

Flowering.   July, August, and the early part of September.

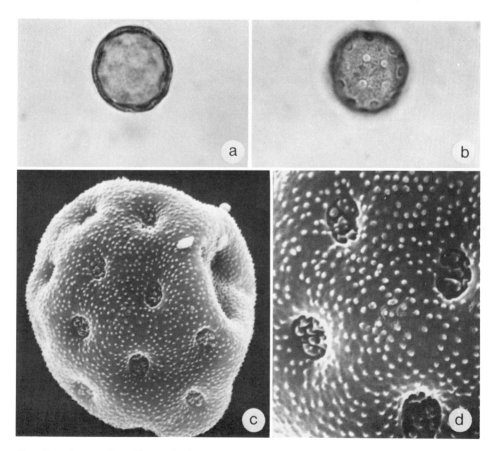

Fig. 42. *Amaranthus albus. a–b*, light microscope, × 1000, *a*, equatorial section, *b*, surface; *c–d*, SEM, *c*, × 1750, *d*, × 7350.

Native to. North America.

Distribution. British Columbia to Newfoundland in open disturbed and waste places.

Notes. At peak flowering, which is the same as the other *Amaranthus* spp., the pollen can cause hay fever.

AMARANTHACEAE

Amaranthus retroflexus L. Redroot pigweed, green amaranth, redroot, rough pigweed.

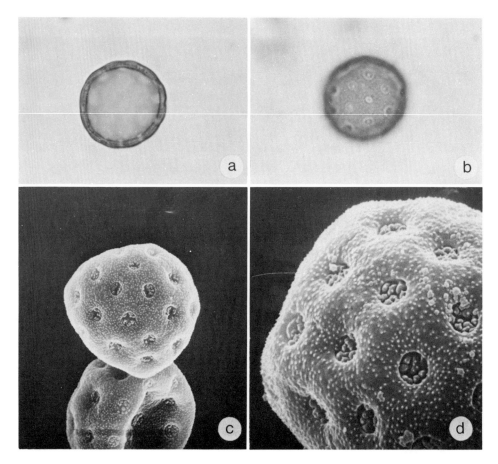

Fig. 43. *Amaranthus retroflexus. a–b*, light microscope, × 1000, *a*, equatorial section, *b*, surface; *c–d*, SEM, *c*, × 1600, *d*, × 3900.

Description. Grains periporate, spheroidal, 23–29 μm, av 26 μm in diam; pores 40–55, av 47, 2 μm in diam, circular, somewhat depressed or intruding with a covered membrane of an irregular number of granules; sculpturing microechinate or scabrate; ektexine and endexine of equal thickness, wall 2 μm thick; structure tectate.

SEM: Grains similar to those of *Salsola pestifer* L. except for smaller pore openings.

Flowering. Late July and August.

Native to. Europe and Asia.

Distribution. Open disturbed habitats from British Columbia to Nova Scotia.

Notes. It is difficult to separate the grains of the amaranth family from those in the goosefoot family (Chenopodiaceae). In areas where redroot pigweed is abundant its pollen probably causes hay fever during the latter part of the summer. Vaughan and Black (1948) stated that in general the amaranth group, pollinates from July to September and is a factor of some importance in the United States in causing hay fever. It could also be troublesome where abundant in Canada.

BETULACEAE

Alnus crispa (Ait.) Pursh.   Green alder, mountain alder.

Description. Grains stephanoporate, spheroidal to irregular in shape, isopolar, (3, 4–) 5 (–6) pores equally distributed on the equators; in polar view 19–26 μm, av 22 μm in diam; in meridional view slightly elliptical, 17–23 μm, av 19 μm wide; pores strongly aspidate giving the grains a fluted or scalloped appearance (in polar view), 1.5–2.0 μm in diam; sculpturing scabrate; ektexine and endexine of equal thickness, wall 3.5–4.5 μm thick; structure tectate.

SEM: Sculpturing scabrate, pores aspidate.

Flowering. June and July.

Native to. North America.

Distribution. The Yukon and Northwest Territories and all provinces except British Columbia.

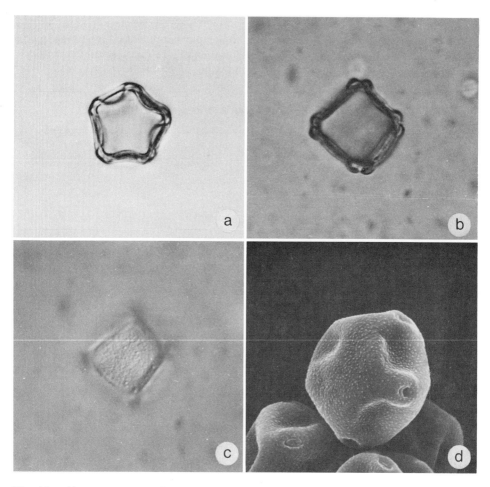

Fig. 44. *Alnus crispa*. a–c, light microscope, × 1000, a, equatorial section, 5 pores, b, equatorial section, 4 pores, c, surface under interference contrast; d, SEM, × 1600.

Notes. Soft alder, *Alnus crispa* var. *mollis* Fern., which occurs from Ontario to Newfoundland, flowers in May and June. The pollen of the soft alder is very similar to that of the green alder. Alder pollen causes hay fever when shed in large amounts (Vaughan and Black 1948).

BETULACEAE

Alnus rugosa (Du Roi) Spreng.   Speckled alder, gray alder, hoary alder.

Description.   Grains stephanoporate, spheroidal to irregular in shape, isopolar, (3–) 4–5 (–6) pores, equally distributed near the equators; in polar view 20–27 μm, av 24 μm in diam; in meridional view 16–26 μm, av 21 μm wide; pores strongly aspidate giving the grain a ridged, fluted, or scalloped appearance (in polar view), 2–4 μm in diam; sculpturing scabrate rugulate; ektexine and endexine of equal thickness, wall less than 2 μm thick; structure tectate.

SEM: Grains similar to those of *Alnus crispa*.

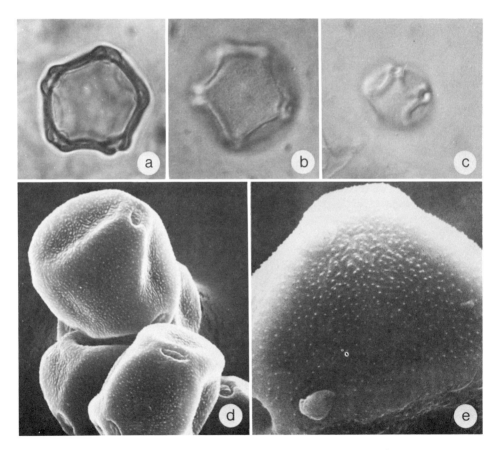

Fig. 45.   *Alnus rugosa.* a–c, light microscope, a–b, × 1000, c, × 400. a, equatorial section, b, surface under interference contrast, c, arcus and pores under interference contrast; d–e, SEM, d, × 1640, e, × 3780.

Flowering. March to June depending on regional climatic conditions.

Native to. Eastern North America.

Distribution. Saskatchewan to Newfoundland.

Notes. Grains of *Alnus rugosa* are very similar to those of *A. crispa* except their pore openings are larger when seen in polar view.

The pollen of the *Alnus oregona* Nutt., red alder, *A. tenuifolia* Nutt., mountain alder, and *A. sinuata* (Reg.) Rydb., Sitka alder, which occur in British Columbia, were not examined under the SEM. It is possible that the pollen of red alder, mountain alder, Sitka alder, and speckled alder can cause hay fever when shed in large amounts.

Pollen Key to the Species of *Betula* (Birches)

A. Ektexine and endexine of equal thickness

    B. Grains averaging 30 µm in diam (polar view) .... ***Betula alleghaniensis***

    B. Grains averaging 24 µm in diam (polar view) .... ***B. populifolia***

A. Ektexine at least twice as thick as the endexine

    C. Sculpturing scabrate .... ***B. occidentalis***
                                                                                                 B. minor

    C. Sculpturing rugulate with scabrate elements .... ***B. papyrifera***

BETULACEAE

Betula alleghaniensis Britton   Yellow birch, curly birch.

Description. Grains triporate, occasionally tetraporate; in isopolar view, 26–35 µm, av 30 µm in diam; in meridional view elliptical, 23–30 µm, av 26 µm wide; pores annulate, vestibulate, 3–6 µm in diam, annulus (including pore) up to 10 µm in diam; sculpturing rugulate with scabrate elements; ektexine and endexine of equal thickness, wall 1–2 µm thick; structure tectate.

    SEM: Grains similar to those of *B. occidentalis,* water birch.

Flowering. April and May.

Native to. Central and eastern North America.

Distribution. Western Ontario to Newfoundland, most common in southern Ontario, Quebec and part of the Maritime Provinces.

Notes. The pollen morphology of *B. lenta* L., cherry birch, which is only known in one location in Ontario, is very similar to that of yellow birch.

There appears to be crossed allergic reactivity among all taxa of birch (Vaughan and Black 1948).

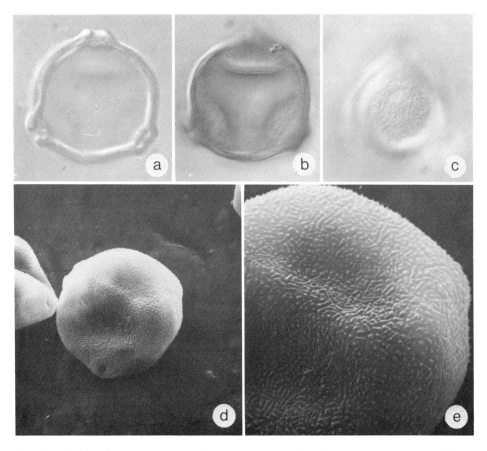

Fig. 46. *Betula alleghaniensis. a-c,* light microscope, (interference contrast), × 1000, *a,* polar view, equatorial section, *b,* arcus, *c,* surface; *d-e,* SEM, *d,* × 1540, *e,* × 3850.

BETULACEAE

Betula minor (Tuckerm.) Fern.   Dwarf white birch, dwarf birch.

Description. Grains triporate, occasionally tetra or pentaporate; in polar view circular, 25–31 μm, av 28 μm in diam; in meridional view slightly elliptical, 24–30 μm, av 28 μm wide; pores annulate, vestibulate, 3.0–4.5 μm in diam, annulus (including pore) up to 6 μm in diam; sculpturing scabrate elements; ektexine twice as thick as the endexine, wall 2.0 μm thick; structure tectate.

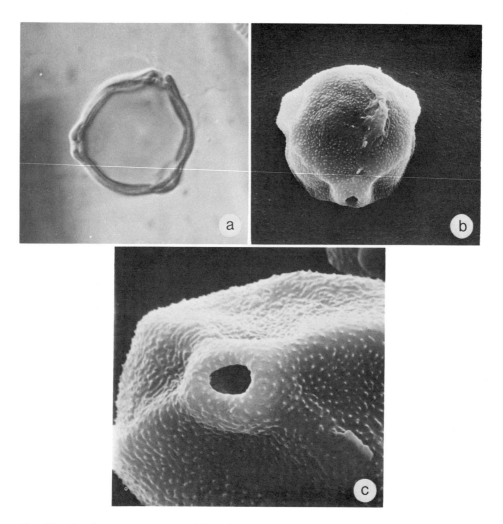

Fig. 47. *Betula minor. a*, equatorial section, light microscope, (interference contrast), × 1000; *b–c*, SEM, *b*, × 1820, *c*, × 4550.

SEM: Grains with protruding pores, an arcus; sculpturing microscabrate processes evenly distributed on a smooth exine.

Flowering. June and July.

Native to. Eastern North America.

Distribution. Primarily in Quebec and Newfoundland.

Notes. The pollen grains of *B. glandulosa* Mich., dwarf birch, and *B. pumila* L., swamp birch, are similar to those of *B. minor*. There appears to be crossed allergic reactivity among all taxa of birch (Vaughan and Black 1948).

BETULACEAE

Betula occidentalis Hook. Water birch, black birch.

Description. Grains triporate, occasionally tetraporate, isopolar; in polar view circular, 24–29 μm, av 26 μm in diam; in meridional view elliptical, 20–24 μm, av 22 μm wide; pores annulate with a prominent vestibulum, 3.0–4.5 μm in diam, annulus (including pore) up to 8 μm in diam; sculpturing scabrate elements; ektexine twice as thick as the endexine, wall 1.5–2.0 μm thick; structure tectate.

SEM: Sculpturing scabrate, elements on striate ridges.

Flowering. May and June.

Native to. Central and western North America.

Distribution. British Columbia to Manitoba and the Northwest Territories.

Notes. The pollen grains are similar to those of *B. papyrifera*, white birch. There appears to be crossed allergic reactivity among all taxa of birch (Vaughan and Black 1948).

(Fig. 48 overleaf)

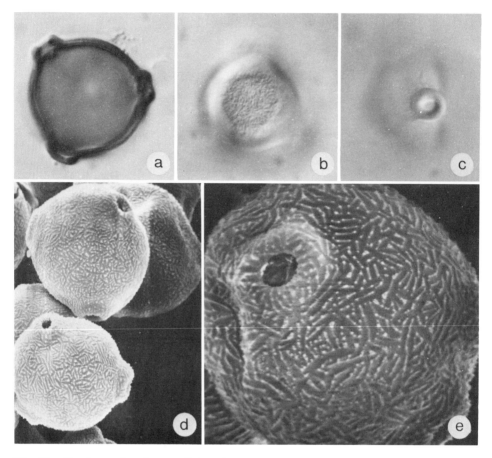

Fig. 48. *Betula occidentalis*. *a–c*, light microscope (interference contrast), × 1000, *a*, polar view, equatorial section, *b*, surface, *c*, pore; *d–e*, SEM, *d*, × 1540, *e*, × 3850.

BETULACEAE

Betula papyrifera Marsh.   White birch, paper birch, canoe birch.

Description. Grains triporate, rarely tetraporate, isopolar; in polar view circular, 22–32 μm, av 28 μm in diam; in meridional view elliptical, 19–28 μm, av 24 μm wide; pores annulate, vestibulate, 3–5 μm in diam, annulus (including pore) up to 10 μm in diam (in some pollen the annulus splits giving the impression of colporate grains); sculpturing rugulate with scabrate elements; ektexine more than twice as thick as the endexine, wall 1.0–1.5 μm thick; structure tectate.

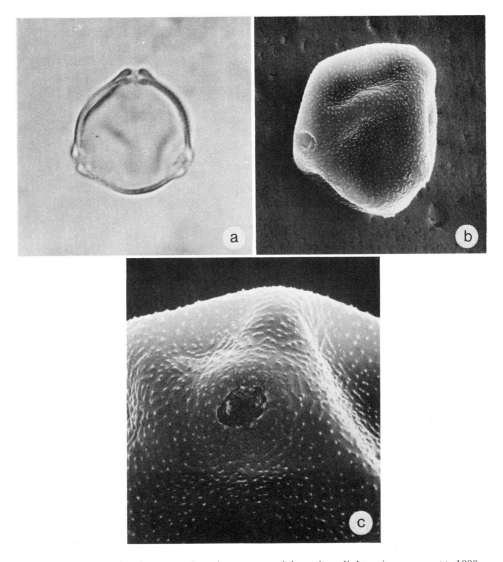

Fig. 49. *Betula papyrifera.* a, polar view, equatorial section, light microscope, × 1000; b-c, SEM, b, × 1750, c, × 4375.

SEM: Sculpturing scabrate, elements on a smooth undulating exine surface.

Flowering. April and May.

Native to. North America.

Distribution. All provinces and the Yukon, abundant in many areas especially in Eastern Canada.

Notes. The pollen morphology of the *B. neolaskana* Sarg., Alaska birch, is very similar to that of the white birch. When shed in large amounts, the pollen of *B. papyrifera* is known to cause hay fever.

BETULACEAE

Betula populifolia Marsh.   Gray birch, wire birch.

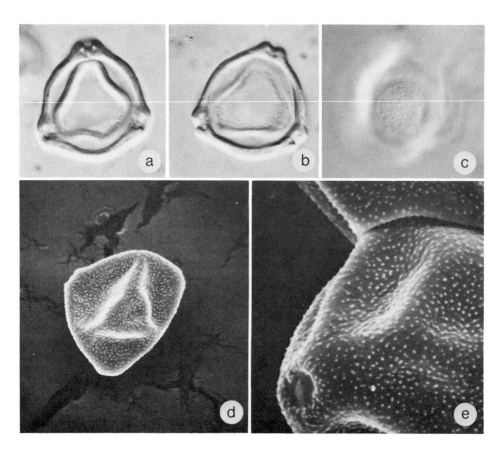

Fig. 50. *Betula populifolia.* a–c, light microscope (interference contrast), × 1000, a–b, polar view, equatorial section, c, surface; d–e, SEM, d, × 1540, e, × 3850.

Description. Grains triporate, rarely tetraporate, isopolar; in polar view circular, 21–29 μm, av 24 μm in diam; in meridional view elliptical, 18–21 μm, av 20 μm wide; pores annulate, occasionally a few grains with nonannulate pores, vestibulate, 2.5–3.5 μm in diam, annulus (including pores) up to 7 μm in diam; sculpturing scabrate elements; ektexine and endexine of equal thickness, wall 2 μm thick; structure tectate.

SEM: Grains similar to those of *Betula papyrifera.*

Flowering. April and May.

Native to. Eastern North America.

Distribution. Eastern Ontario to Nova Scotia, fairly common in southern Quebec and parts of the Maritime Provinces.

Notes. The pollen is similar to that of *B. occidentalis.* There appears to be crossed allergic reactivity among all taxa of birch (Vaughan and Black 1948).

BETULACEAE

Corylus cornuta Marsh. Beaked hazel, horned hazelnut.

Description. Grains triporate; in polar view spheroidal to triangular, 21–27 μm, av 23 μm in diam; in meridional view elliptical, 18–21 μm, av 19.5 μm wide; pores 2–3 μm in diam, circular, sometimes operculate, no distinctive tumescence of ektexine in area of the pore; sculpturing scabrate rugulate; ektexine and endexine of equal thickness, endexine granular in pore area, wall 1.0–1.5 μm thick; structure tectate.

SEM: Sculpturing on grains similar to that of *Betula papyrifera.*

Flowering. April.

Native to. North America.

Distribution. British Columbia to Newfoundland.

Notes. Pollen in large amounts may cause hay fever.

(Fig. 51 overleaf)

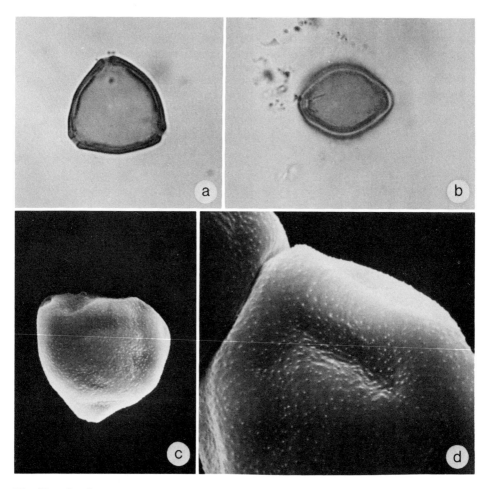

Fig. 51. *Corylus cornuta*. a-b, light microscope, × 1000, a, polar view, equatorial section, b, equatorial view showing shape and position of pores; c-d, SEM, c, × 1600, d, × 4000.

BETULACEAE

Ostrya virginiana (Mill.) K. Koch   Hop-hornbeam, ironwood, rough barked ironwood, hornbeam.

Description. Grains triporate, occasionally tetraporate; in polar view triporate grains circular to triangular, 22–27 μm, av 25 μm in diam; in meridional view 19–24 μm, av 22 μm wide; pores equally distributed around the equator of grain, 2–3 μm in diam, annulate, ektexine slightly tumescent in pore area; sculpturing scabrate; ektexine and endexine of equal thickness, wall 1.0–1.5 μm thick; structure tectate.

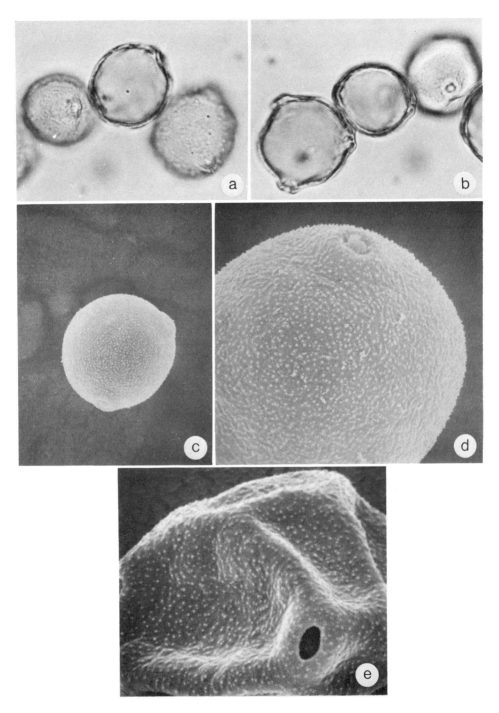

Fig. 52. *Ostrya virginiana. a–b*, light microscope, × 1000; *c–e*, SEM, *c*, × 1400, *d*, × 3500, *e*, × 3850.

SEM: Sculpturing of grains similar to that of *Betula papyrifera*.

Flowering. April and May.

Native to. Eastern North America.

Distribution. Eastern Manitoba to Nova Scotia.

Notes. Vaughan and Black (1948) mentioned that the pollen of ironwood is not important in causing hay fever, but positive skin reactions have been observed in preliminary tests.

Carpinus caroliniana Walt., blue-beech, has pollen grains very similar to those of ironwood and several of the birches, particularly when blue-beech pollen has four pores (Wodehouse 1935). Where common, blue-beech sheds large quantities of pollen in the early spring and can cause hay fever. This tree occurs sporadically in southern Ontario and southwestern Quebec.

CANNABINACEAE

Cannabis sativa L. Marijuana, hemp.

Description. Grains triporate, occasionally tetraporate; suboblate in shape; in polar view 22–26 µm, av 24 µm in diam; in meridional view 23–27 µm, av 25 µm wide; pores aspidate, circular in outline, 1–2 µm in diam, annulus 1.0–1.5 µm thick, without an operculum (some unacetolyzed grains with 1, 2, or 3 granules within the pore); sculpturing unevenly distributed microechinate or scabrate processes; ektexine twice as thick as the endexine, wall 1.0 µm thick; structure tectate.

SEM: Grains with evenly distributed microechinate particles of equal size over the whole grain.

Flowering. July, August, to the middle of September.

Native to. Asia.

Distribution. Rare in southern British Columbia, Alberta, and Manitoba; some in disturbed areas of southern Ontario and in Quebec as far east as the Gaspé.

Notes. In 1972 marijuana was grown at the Central Experimental Farm at Ottawa for experimental studies. Pollen shed from this taxon was caught in large amounts on exposed slides during the flowering period. Several workers in the area developed hay fever symptoms at the time of flowering. They did not have this experience before or after the experiment. Feinberg (1946)

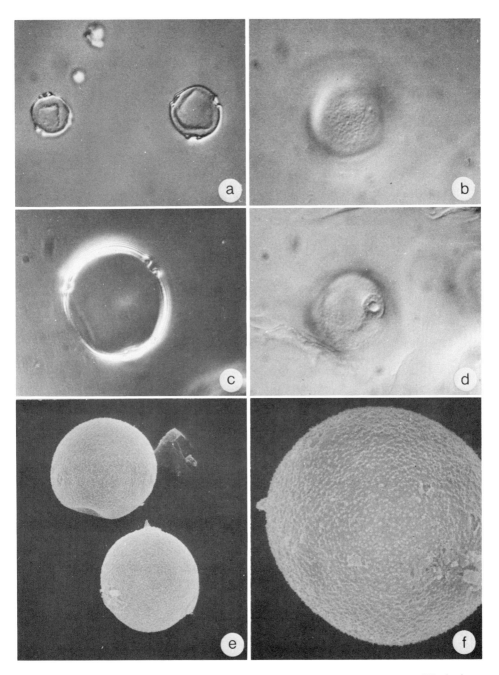

Fig. 53. *Cannabis sativa. a–d*, light microscope (interference contrast), *a*, × 400, *b–d*, × 1000; *e–f*, SEM, *e*, × 1400, *f*, × 3500.

mentioned that marijuana is a good example of a plant that is capable of causing a great deal of hay fever, but the areas of heavy distribution are strictly local.

Pollen of the common hop, *Humulus lupulus* L., under the light and SEM microscopes is similar to that of *Cannabis sativa*. Common hop flowers in July and August and is widely cultivated in British Columbia, southern Ontario, Quebec, and the Maritime Provinces. Hop pollen has been mentioned as a possible cause of inhalant allergy, but clinical confirmation is lacking (Feinberg 1946).

CAPRIFOLIACEAE

Sambucus canadensis L.   Common elder, elderberry.

Description.   Grains tricolporate, prolate, rarely circular; in meridional view, 13–19 μm wide, 19–24 μm, av 16–22 μm long; polar area index 0.21, pores not distinct and appear as a protuberance on the grain equator nestled within the colpi; sculpturing microreticulate over the whole grain with a finer reticulate pattern near the colpi edges; ektexine slightly thicker than the endexine, wall 1 μm thick; structure tectate.

SEM: Sculpturing microreticulate, lumina irregular, 0.3–0.8 μm in diam; muri supported by micropilae with the top elevation of the muri rounded.

Flowering.   July and August.

Native to.   Eastern North America.

Distribution.   Ontario to Nova Scotia.

Notes.   The pollen of *Sambucus pubens* Michx. (ranges from British Columbia to Newfoundland) is somewhat more circular than the grains of *S. canadensis*. *S. pubens* flowers from May to July. A few pollen grains from *Sambucus* spp. were caught on exposed slides in the Atlantic Provinces. There is no information that the pollen of *Sambucus* spp. causes hay fever.

Although the concentrations of airborne entomophilous pollen do not closely approach the concentrations of pollen found close to many anemophilous plants, the following data prepared by Ogden et al. (1975) of the University of the State of New York, Albany, New York, USA, indicate significant amounts with the battery-powered rotoslide samplers. All figures represent the maximum recorded at 1.5 m above ground and at least 1 m from nearest flowers: *Sambucus canadensis,* grains 125 000/m^3; *Sorbaria sorbifolia* A. Br. (Rosaceae), grains 3736/m^3; *Spiraea vanhouttei* Zabel (Rosaceae), grains 1272/m^3; and several others with grains below 400/m^3. Although Ogden et al. (1975) sampled

125 species, these data are preliminary and might be greatly changed through further sampling. Several of these entomophilous taxa may play an important role in causing hay fever.

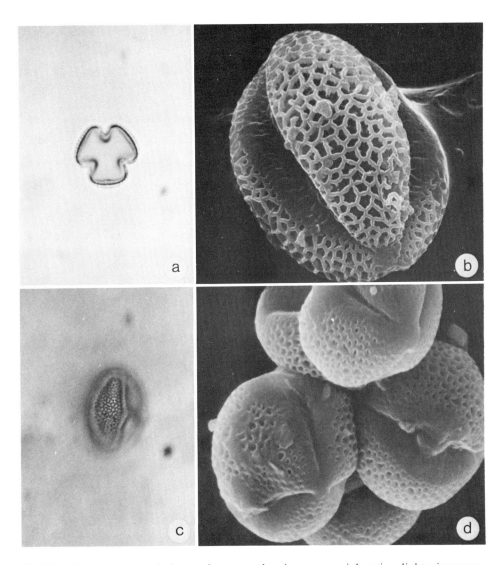

Fig. 54. *Sambucus* spp. *a–b*, *S. canadensis*, *a*, polar view, equatorial section, light microscope, × 1000, *b*, SEM, × 3850. *c–d*, *S. pubens*, *c*, equatorial surface, light microscope, × 1000, *d*, SEM, × 2000.

CARYOPHYLLACEAE

Arenaria serpyllifolia L.   Thyme-leaved sandwort, sandwort.

Description.   Grains periporate, spheroidal, 25–31 μm, av 28 μm in diam; pores about 30 over the whole grain, 5–7 μm in diam with slight annulate thickenings and granular pore membranes over the surface; sculpturing not clearly defined; endexine thicker than the ektexine, wall 3–4 μm thick; structure tectate, baculate.

SEM: Sculpturing microechinate with spines evenly distributed over the exine surface; small pits or microperforations in some areas of the exine.

Flowering.   April to September.

Native to.   Europe.

Distribution.   British Columbia to Nova Scotia.

Notes.   Occasionally a few pollen grains from this family are caught on exposed slides. Most plants in this family are not considered to be wind-pollinated. The pollen of most species are similar to those in the families Chenopodiaceae and Plantaginaceae. There is no specific information that pollen from the Caryophyllaceae causes hay fever.

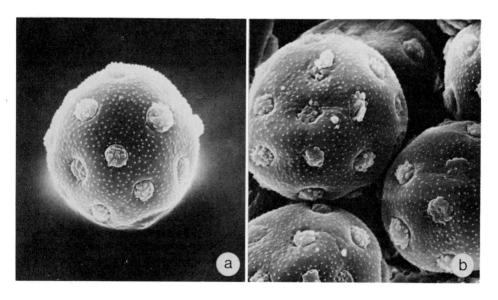

Fig. 55.   *Arenaria serphyllifolia. a–b*, SEM, × 2400.

Pollen Key to Nine Species in the Chenopodiaceae (Lamb's-Quarters Family)

A. Grains with annulate pores

    B. Grains with 25 or less pores; grains not exceeding 22 μm in overall diam .... ***Sarcobatus vermiculatus***

    B. Grains with 100 or more pores; grains exceeding 22 μm in overall diam .... ***Atriplex subspicata***

A. Grains without annulate pores

    C. Pores positioned more or less even with the exine surface, not recessed or in depressions

        D. Pores normally exceeding 100 in number

            E. Pore diam 3 μm or more

                F. Wall thickness 2 μm or more .... ***Chenopodium berlandieri* ssp. *zschackei***

                F. Wall thickness less than 2 μm .... ***Kochia scoparia***

            E. Pore diam 2.5 μm or less

                G. Grains normally with more than 125 pores ...... ***Chenopodium album***

                G. Grains normally with less than 115 pores ...... ***Axyris amaranthoides***

        D. Pores normally less than 100 in number

            H. Wall thickness more than 3 μm .... ***Eurotia lanta***

            H. Wall thickness 2 μm or less .... ***Suaeda maritima***

    C. Pores not evenly positioned with the exine surface, recessed or in depressions .... ***Salsola pestifer***

CHENOPODIACEAE

Axyris amaranthoides L.   Russian pigweed.

Description. Grains periporate, spheroidal, 24–28 μm, av 26 μm in diam; pores 2.0–2.5 μm, av 2.2 μm wide, 100–115, av 110 pores over the whole grain; ektexine and endexine about 1.5 μm thick; granular particles scattered over the whole grain and the pore wall.

    SEM: Small granular particles scattered over the whole grain and over the pore wall; margins of pores entire.

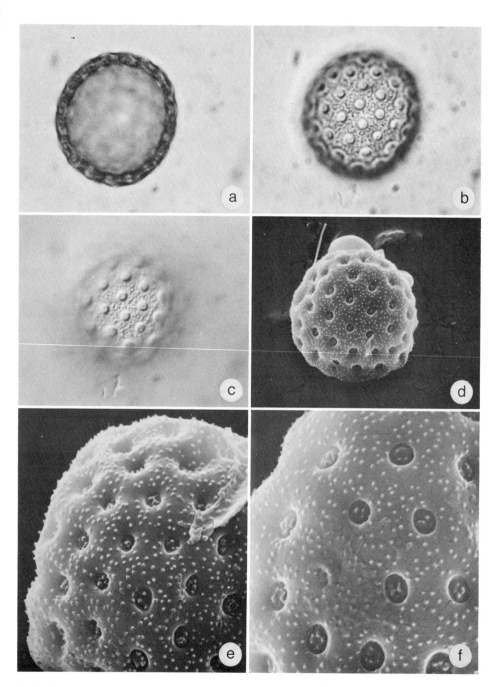

Fig. 56. *Axyris amaranthoides*. a–c, light microscope, × 1000, a, optical section, b, surface and pores, c, surface and pores under interference contrast; d–f, SEM, d, × 1470, e, × 3780, f, × 4270.

Flowering. Begins around the latter part of July and continues through August, with the peak period the first 2 weeks of August.

Native to. Siberia.

Distribution. All provinces except Newfoundland, abundant in the Prairie Provinces, rare in Eastern Canada.

Notes. Because this is mostly a wind-pollinated herb, it sheds large amounts of pollen and where abundant it is an important factor in causing hay fever.

CHENOPODIACEAE

Atriplex subspicata (Nutt.) Rydb.   Orach, saltbrush, saline orach.

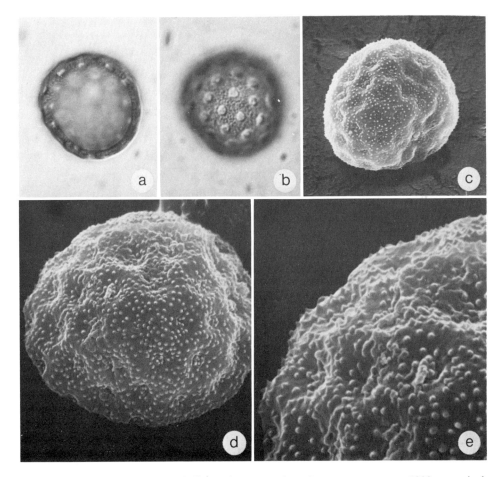

Fig. 57. *Atriplex subspicata. a–b,* light microscope, interference contrast, × 1000, *a,* optical section, *b,* surface; *c–e,* SEM, *c,* × 1600, *d,* × 3700, *e,* × 7350.

Description. Grains periporate, spheroidal, 24–29 µm, av 26 µm in diam; pores circular, slightly annulate, 2 µm in diam with a granular membrane, 90–120 in number; sculpturing of scabrate elements, not very distinctive; ektexine and endexine of equal thickness, wall 1.5–2.0 µm thick; structure tectate.

SEM: Sculpturing on surface, evenly spaced, acute, microechinate processes; pores recessed producing an undulating surface.

Flowering. Mid-July to the early part of September.

Native to. North America.

Distribution. All provinces and the Northwest Territories, particularly common and widespread in the saline areas of the Prairie Provinces.

Notes. Wodehouse (1971) stated that the pollen of all species in the genus *Atriplex* can cause hay fever. In areas where abundant, especially in Western Canada, the pollen probably causes considerable hay fever.

CHENOPODIACEAE

Chenopodium album L.   Lamb's-quarters, fat hen, pigweed, white goosefoot.

Description. Grains periporate, spheroidal, 23–34 µm, av 28 µm in diam; pores circular, 2.0–2.5 µm in diam with a granular membrane, (80–) 90–140 (–150) in number; sculpturing a smooth surface flecked with regularly spaced microgranules; ektexine thicker than the endexine, wall 2.0–2.5 µm thick, ektexine and tectum joined by short baculate rods; structure tectate.

SEM: Sculpturing microechinate, elements evenly distributed over the smooth surface; granular membrane over the pore area.

Flowering. July, August, and September.

Native to. Europe and Asia.

Distribution. All provinces, abundant in all agricultural areas.

Notes. Although lamb's-quarters is entirely wind-pollinated, the plants do not shed large amounts of pollen. However, where there are many plants the pollen may occasionally be a real factor in causing hay fever (Wodehouse 1971).

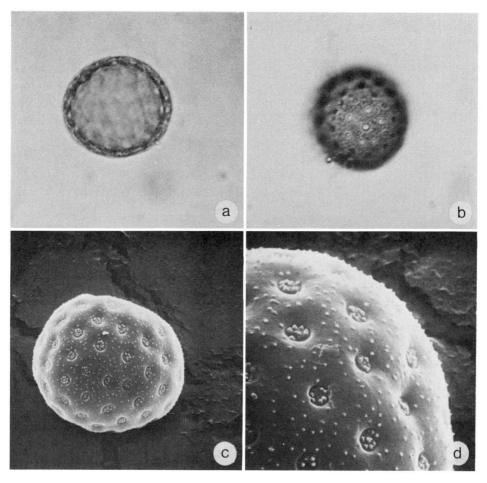

Fig. 58. *Chenopodium album.* a–b, light microscope, × 1000, a, optical section, b, surface; c–d, SEM, c, × 1575, d, × 3920.

CHENOPODIACEAE

Chenopodium berlandieri Moq. ssp. *zschackei* (Murr.) Zobel. Net-seeded lamb's-quarters, lamb's-quarters, goosefoot.

Description. Grains periporate, spheroidal, 24–30 μm, av 26 μm in diam; pores 75–110, circular with a granular membrane, 3–4 μm in diam; sculpturing, wall thickness, and structure similar to the grains of *C. album.*

SEM: Sculpturing similar to the grains of *C. album* except for somewhat broader based microechinate spines.

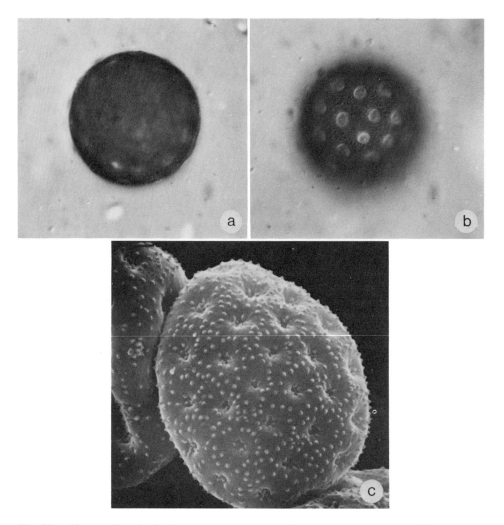

Fig. 59. *Chenopodium berlandieri* ssp. *zschackei. a–b*, light microscope, × 1000, *a*, optical section, *b*, surface; *c*, SEM, × 3920.

Flowering. Late July, August, and September.

Native to. North America.

Distribution. British Columbia to Ontario, probably more abundant than lamb's-quarters in the Prairie Provinces (Frankton and Mulligan 1970).

Notes. Where abundant the pollen may cause hay fever.

CHENOPODIACEAE

Eurotia lanata (Pursh) Moq. Winter fat.

Description. Grains periporate, spheroidal, 23–28 μm, av 25.5 μm in diam; pores 3.0–4.0 μm, av 3.5 μm in diam, 55–65, av 60 pores over the whole grain; ektexine and endexine about 1.5 μm thick; many granular particles over the whole grain and the pore wall.

SEM: Many small granular particles over the whole grain and over the pore, margins of pore wall irregular.

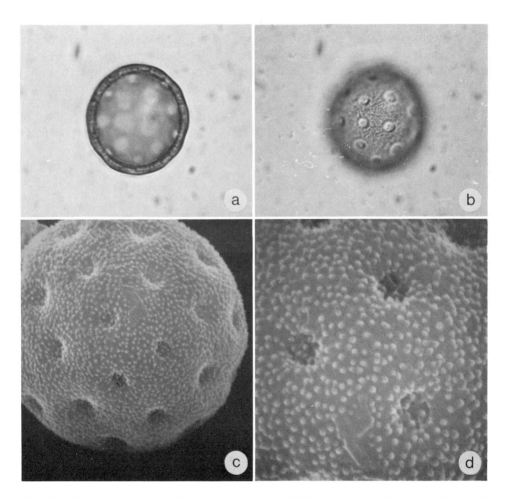

Fig. 60. *Eurotia lantana. a–b*, light microscope, × 1000, *a*, optical section, *b*, surface; *c–d*, SEM, *c*, × 4200, *d*, × 8400.

Flowering. In Alberta and Saskatchewan the first week of July to the early part of September, maximum flowering the latter part of July and the first week of August.

Native to. North America.

Distribution. Widely scattered in southern Alberta and Saskatchewan, known in the Yukon.

Notes. The pollen shed in large amounts may cause hay fever.

CHENOPODIACEAE

Kochia scoparia (L.) Schrad. Kochia, burning bush, summer cypress, fire bush, belvedere.

Description. Grains periporate, spheroidal, 28–33 μm, av 29.5 μm in diam; pores 2.5–3.5 μm, av 2.7 μm wide, 100–130, av 120 pores over the whole grain; ektexine and endexine about 1.5 μm thick; granular particles over the whole grain and the pore membrane.

SEM: Small granular particles over the grain and the pore membrane; margins of pore entire.

Flowering. Latter part of July and through August and the early part of September, with a peak flowering period the first 3 weeks of August.

Native to. Eurasia.

Distribution. All provinces from British Columbia to Nova Scotia except New Brunswick, common in the Okanagan Valley of British Columbia and in the Prairie Provinces, a problem in cultivated fields and waste places around towns and larger centers, rare in Eastern Canada.

Notes. Kochia sheds large quantities of pollen and where common can be counted among the worst hay-fever weeds.

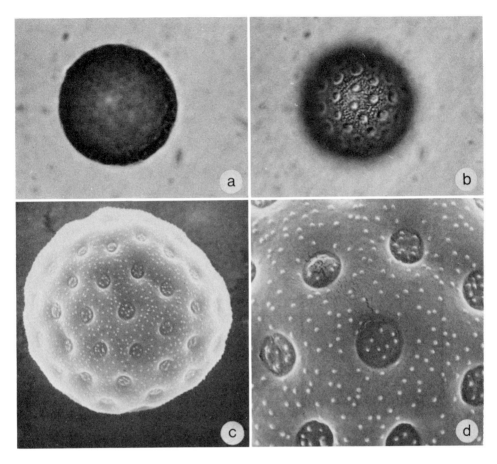

Fig. 61. *Kochia scoparia.* a-b, light microscope, × 1000, a, optical section, b, surface; c-d, SEM, c, × 1700, d, × 4200.

CHENOPODIACEAE

Salsola pestifer A. Nels.   Russian thistle, Russian cactus, Russian tumbleweed, saltwort.

Description. Grains periporate, spheroidal, 26.0–29.0 μm, av 26.0 μm in diam; pores 3.5–4.5 μm, av 3.7 μm in diam; (40–) 50–65 (–75) pores over the whole grain, sunken or depressed; ektexine and endexine about 1.5 μm thick; granular particles scattered over the whole grain and the pore membrane.

SEM: Small granular particles over the whole grain and in more abundance over the pore membrane; margins of pore slightly irregular.

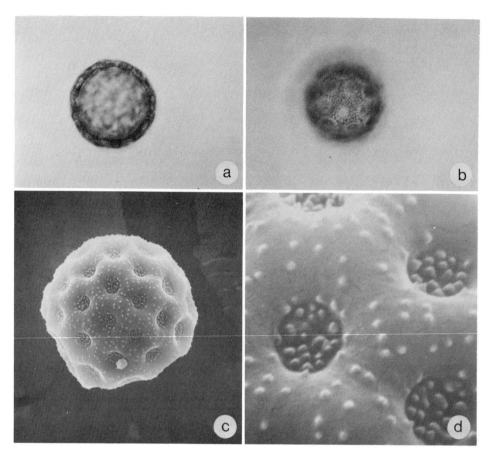

Fig. 62. *Salsola pestifer. a–b*, light microscope, × 1000, *a*, optical section, *b*, surface; *c–d*, SEM, *c*, × 1700, *d*, × 8400.

Flowering. The latter part of July to the end of September, with a peak flowering period from the middle to the end of August.

Native to. Eurasia.

Distribution. Every province except Newfoundland, common in the dry interior of southern British Columbia and in the Prairie Provinces, rare in Eastern Canada.

Notes. This species sheds considerable pollen in Western Canada and is probably the most important member of this group to cause hay fever in late summer and early fall.

CHENOPODIACEAE

Sarcobatus vermiculatus (Hook.) Torr.   Greasewood.

Description.   Grains periporate, spheroidal, 20.5–21.5 μm, av 21.0 μm in diam; pores annulate, 2.5–3.0 μm, av 2.6 μm in diam, 25 pores over the whole grain; ektexine and endexine about 1 μm thick; granular particles over the whole grain and the pore membrane.

SEM: Small granular particles over the whole grain and the pore membrane; margins of pore mostly entire.

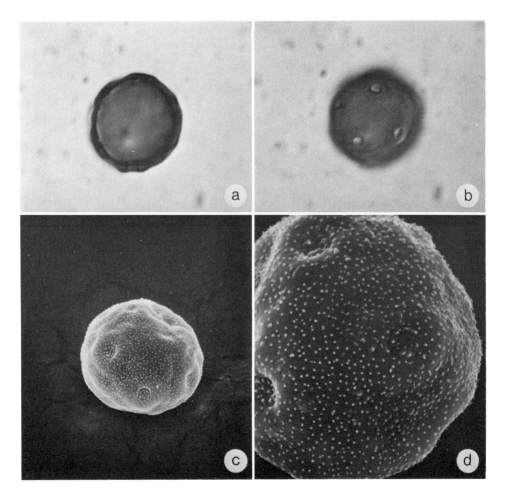

Fig. 63.   *Sarcobatus vermiculatus. a–b*, light microscope, × 1000, *a*, optical section, *b*, surface; *c–d*, SEM, *c*, × 1600, *d*, × 3700.

Flowering. About the latter part of July to early October, with a peak flowering period the last 2 weeks of August and the first week of September.

Native to. North America.

Distribution. Scattered and common in the dry areas of southern Alberta and Saskatchewan, known in British Columbia.

Notes. When shed in large amounts, the pollen can cause hay fever.

CHENOPODIACEAE

Suaeda maritima (L.) Dumort.   Sea-blite, seep-weed.

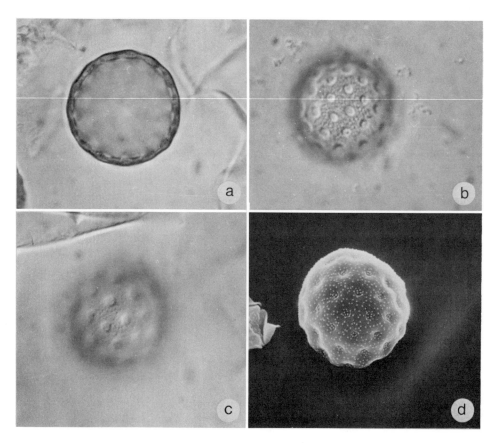

Fig. 64. *Suaeda maritima. a–c*, light microscope, × 1000, *a*, optical section, *b*, surface, *c*, surface under interference contrast; *d*, SEM, × 1400.

Description. Grains periporate, spheroidal, 24–30 μm, av 27 μm in diam; pores 60–100, circular with a granular membrane, 2 μm in diam; sculpturing similar to the pollen of *Chenopodium album*; ektexine and endexine not distinct, wall 1.5 μm thick; structure tectate.

SEM: Sculpturing acute microechinate processes, not densely spaced over the entire grain surface; 4–6 processes on the pore membrane.

Flowering. July, August, and September.

Native to. Eastern North America.

Distribution. Common along the eastern coastline.

Notes. When shed in large amounts, pollen of *Suaeda* spp. may cause hay fever.

Pollen Key to the Ragweeds, *Ambrosia* spp.; Other Ragweed Relatives, *Iva* spp., *Xanthium* spp., and Goldenrod, *Solidago Canadensis*

A. Grains spheroidal to slightly oblate in meridional view

    B. Grains 18–24 μm in overall dimension; wall more than 3 μm thick   ***Iva axillaris***

    B. Grains 24–27 μm in overall dimension; wall less than 2 μm thick

        C. Spines small, 2 μm or less long

            D. Spines under the SEM ca. 60 on half the pollen surface .... ***Xanthium spinosum***

            D. Spines under the SEM ca. 90 on half of the pollen surface .... ***X. strumarium***

        C. Spines large, 3–5 μm long .... ***Solidago canadensis***

A. Grains oblate, rarely spheroidal in meridional view

    E. Colpi 5–12 μm long

        F. Colpi more than 10 μm long .... ***Iva xanthifolia***

        F. Colpi 5–8 μm long .... ***Ambrosia psilostachya***

    E. Colpi 2.5–4.5 μm long

        G. Wall about 2 μm thick

H. Polar area index 0.2 .... *Iva frutescens*

H. Polar area index 0.6 .... *Ambrosia chamissonis*

G. Wall less than 2 μm thick

    I. Spines under the SEM ca. 70–75 on half of the pollen surface

        J. Species only known in Alberta and Saskatchewan ....
.... *Ambrosia acanthicarpa*

        J. Species rare in Western Canada, common in Eastern Canada
.... *A. artemisiifolia*

    I. Spines under the SEM ca. 60–65 on half of the pollen surface
.... *A. trifida*

COMPOSITAE

Ambrosia acanthicarpa Hook.   Bur-ragweed, false ragweed.

Description. Grains tricolporate, occasionally tetracolporate (ratio 3:1), polar area index 0.6; in polar view spheroidal, 20–21 μm, av 20.5 μm in diam; in meridional view oblate, 20.0–22.5 μm wide × 17.5–20.0 μm long; sculpturing echinate; ektexine and endexine about 1 μm thick; structure tectate.

SEM: Colpi 3.0–4.5 μm long, enclosed germ pores 1.5–2.0 μm long; spines about 1 μm long, about 3 μm apart, tips sharply pointed.

Flowering. Begins around July 20, reaching its peak in early August, and continuing until the middle of September.

Native to. North America.

Distribution. Small scattered patches in southern Alberta and Saskatchewan in the sandy plains and valleys.

Notes. It is not known as a weed in the cultivated fields of Alberta and Saskatchewan. Wodehouse (1971) stated that in many places in the United States, principally west of the Mississippi River, it is the commonest weed of cultivated fields. Because the overall pollen count from the ragweeds and their relatives is relatively low in Alberta and Saskatchewan, it is doubtful if the pollen of bur-ragweed plays much of a role in causing hay fever.

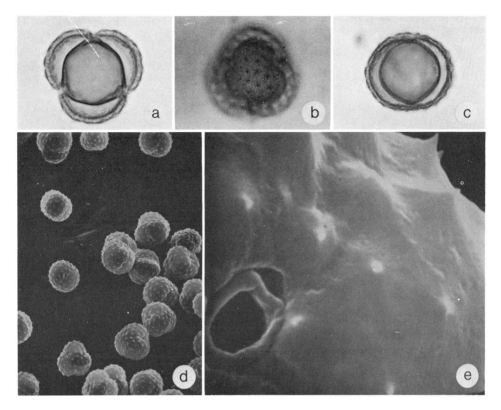

Fig. 65. *Ambrosia acanthicarpa. a–c*, light microscope, × 1000, *a*, polar view, *b*, polar end, *c*, note oblate shape in equatorial view; *d–e*, SEM, *d*, × 500, *e*, × 8750.

COMPOSITAE

Ambrosia artemisiifolia L.   Common ragweed, short ragweed, small ragweed, Roman wormwood.

Description. Grains tricolporate, rarely tetracolporate, polar area index 0.7–0.8; in polar view spheroidal, 19.5–20.5 μm, av 20.0 μm in diam; in meridional view oblate, 19.0–21.5 μm wide × 17.5–19.0 μm long; sculpturing echinate; ektexine and endexine 1.0–1.5 μm thick; structure tectate.

SEM: Colpi 3–5 μm long, enclosed germ pores about 2 μm long; spines 70–75 on half of the pollen surface, 1.5–1.8 μm long, about 3 μm apart, tips sharply pointed.

Flowering. July with the peak period in southern Ontario and Quebec between August 15 and September 15; 2 weeks earlier in southern Manitoba.

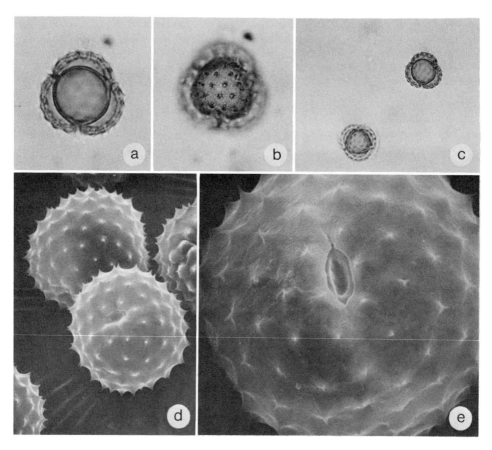

Fig. 66. *Ambrosia artemisiifolia*. a–b, light microscope, × 1000, a, polar view, equatorial section, b, polar end, c, light microscope, × 400; d–e, SEM, d, × 1680, e, × 4000.

Native to. North America.

Distribution. The Northwest Territories and all the provinces.

Northwest Territories: One collection recorded from Fort Smith.

British Columbia: Very rare, collections from the Okanagan Valley and two other locations in suburban areas of southern British Columbia.

Alberta: Relatively rare, small patches in disturbed habitats from Edmonton south to USA (Bassett and Terasmae 1962).

Saskatchewan: Relatively rare, several collections from southeastern part of province.

Manitoba: Rare, scattered small patches in Winnipeg and elsewhere in southern Manitoba especially in disturbed habitats near the Red River.

Ontario: About 65% of the area of heavy ragweed concentration in southern Ontario, northern limit base of Bruce Peninsula, eastern limit Hawkesbury (Bassett 1959).

Quebec: Abundant in the rich lowlands along the Ottawa and St. Lawrence rivers between the Laurentian Mountains in the north and the Appalachian Mountains in the south and east to Quebec City; relatively rare elsewhere.

New Brunswick: Fairly abundant along roadsides for 24 km south of Fredericton towards St. John and in some places between Bathurst and Moncton; relatively rare elsewhere.

Nova Scotia: Fairly abundant in waste places and along roadsides between Wolfville and Annapolis Royal; scattered around Yarmouth and along the eastern coastline from Shag Harbour to Canso and from Truro to Antigonish; rare elsewhere.

Prince Edward Island: Small patches in several urban areas, but not common in any part.

Newfoundland: One specimen collected in a garden near St. John's in 1931, but no other plants reported since.

Notes. Common ragweed is found under a wide variety of soil and moisture conditions in cultivated fields, gardens, vacant lots, waste places, along roadsides, and fencerows. It is a typical after-harvest cover in grainfields (Frankton and Mulligan 1970). The pollen is the most important cause of hay fever in Eastern Canada.

The plant or its pollen may produce a dermatitis in some people who are not necessarily sufferers from hay fever. Dairy products from cows that have grazed on this plant have an objectionable odor and taste (Frankton and Mulligan 1970). A recent account of the biology of common and perennial ragweeds in Canada has been prepared by Bassett and Crompton (1975).

COMPOSITAE

Ambrosia chamissonis (Less.) Greene   Beach sandbur, false ragweed.

Description. Grains tricolporate, occasionally tetracolporate, polar area index 0.6; in polar view spheroidal, 21.5–24.0 μm, av 22.5 μm in diam; in meridional view oblate, 21.0–24.0 μm wide × 18.0–19.5 μm long; sculpturing echinate; ektexine and endexine about 2 μm thick; structure tectate.

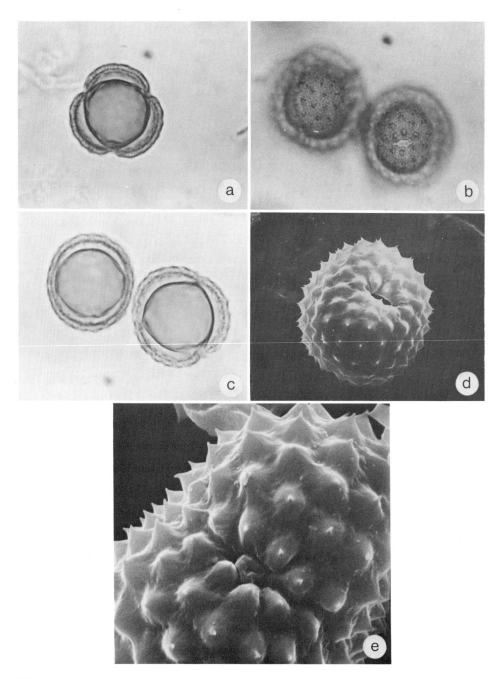

Fig. 67. *Ambrosia chamissonis.* a-c, light microscope, × 1000, a, polar view, equatorial section, b, surface, colpi and pore, c, equatorial and polar section; d-e, SEM, d, × 1750, e, × 4000.

SEM: Colpi 4.5–6.0 μm long; spines about 1 μm long, about 3 μm apart, tips sharply pointed.

Flowering. On Vancouver Island the end of May, peak period in June and July, pollen shed into September and October.

Native to. North America.

Distribution. Widely scattered from southern British Columbia to the Queen Charlotte Islands on the sea beaches and sand dunes; none elsewhere.

Notes. At no time during the summer period was there any large amount of airborne pollen caught on the exposed slides of a gravity sampler set up near the shoreline on Vancouver Island where beach sandbur was growing nearby (Bassett and Crompton 1966).

COMPOSITAE

Ambrosia psilostachya DC.   Perennial ragweed, western ragweed.

Description. Grains tricolporate, rarely tetracolporate, polar area index 0.7; in polar view spheroidal, diam 21–23 μm, av 22 μm; in meridional view oblate, 21.0–23.0 μm wide × 19.0–20.5 μm long; sculpturing echinate; ektexine and endexine about 1.5 μm thick; structure tectate.

SEM: Colpi 5–8 μm long, the enclosed germ pores 4–5 μm long; spines 80–85 on half of pollen surface, 3.0–3.5 μm long, about 3 μm apart, tips sharply pointed.

Flowering. In Western Canada from mid-July to mid-September, with the peak period the first 2 weeks of August.

Native to. North America.

Distribution. All provinces except New Brunswick.

British Columbia: One collection in the southeastern part of province (Bassett and Terasmae 1962).

Alberta: One collection in the southeastern part of the province.

Saskatchewan: Widely scattered south of the Saskatchewan River to USA, mostly in the eastern part of the province.

Manitoba: Greatest concentration in Canada in the southwestern part of the province; very few populations east of the Red River to the Ontario border.

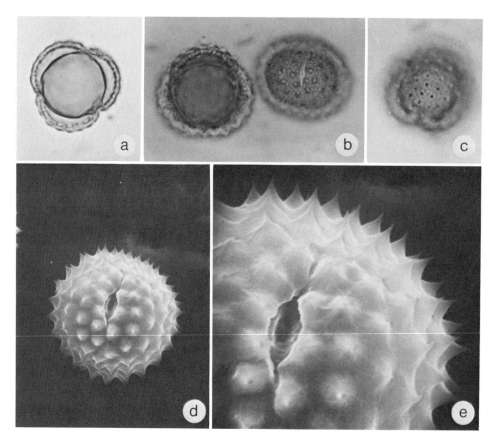

Fig. 68. *Ambrosia psilostachya. a–c*, light microscope, × 1000, *a*, polar view, equatorial section, *b*, colpi and pore, *c*, surface; *d–e*, SEM, *d*, × 1680, *e*, × 4200.

Ontario, Quebec, and Maritime Provinces: Rare.

Notes.   Perennial ragweed occurs in disturbed pasture and abandoned fields, mostly in sandy soils. It is also found along irrigation ditches, roadsides, railroads, vacant lots, and waste places around communities. Large quantities of pollen are shed, but because of the more localized occurrence of the plants and their small size, perennial ragweed is less important as a cause of hay fever than the giant and common ragweeds (Wodehouse 1971).

COMPOSITAE

Ambrosia trifida L.   Giant ragweed, great ragweed, kinghead, tall ragweed.

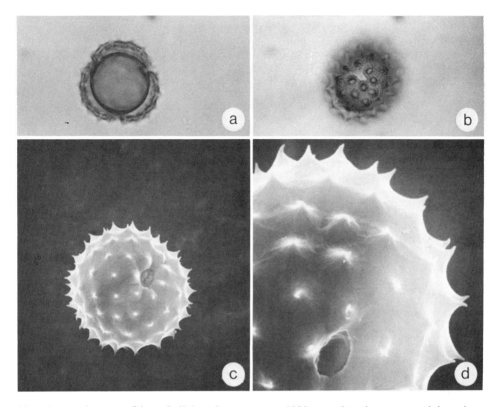

Fig. 69. *Ambrosia trifida*. a–b, light microscope, × 1000, a, polar view, equatorial section, b, surface and colpi; c–d, SEM, c, × 1670, d, × 4200.

Description. Grains tricolporate, rarely tetracolporate, polar area index 0.7–0.8; in polar view spheroidal, 19.5–20.5 μm, av 19.5 μm in diam; in meridional view oblate, 19.5–21.0 μm wide × 17.5–19.0 μm long; sculpturing echinate; ektexine and endexine about 1 μm thick; structure tectate.

SEM: Colpi 3 μm long, the enclosed germ pores 1.5–2.0 μm long; spines 60–65 on half of the pollen surface, 1.5–1.8 μm long, about 3 μm apart, tips sharply pointed.

Flowering. Begins 2 to 3 weeks earlier than common ragweed, reaching its peak near the end of August (Bassett, Holmes, and MacKay 1961).

Native to. North America.

Distribution. Every province except Newfoundland (Bassett and Terasmae 1962).

British Columbia: One collection on Vancouver Island and a second in southern British Columbia near the Alberta border.

Alberta: Rare, only three recorded collections.

Saskatchewan: Several collections, but species rare.

Manitoba: Most abundant of the ragweeds in southern Manitoba along the Red River valley.

Ontario: Although widely distributed, stands very localized except in the extreme southern part of the province.

Quebec: Several patches in southern Quebec, but not as abundant or widespread as the common ragweed.

New Brunswick: Small patches throughout the province, but no abundance in any one area.

Nova Scotia: Relatively rare.

Prince Edward Island: Widely collected, but relatively rare.

Notes. Giant ragweed is found along roadsides, railways, in agricultural fields, and in waste places near towns, usually on rather rich soils. It is sometimes found in undisturbed habitats, in marshes that dry out in summer, or on rich moist soils near streams where it reaches its greatest growth (Frankton and Mulligan 1970). In regions where the giant ragweed outranks the common ragweed in the number of individuals, the former produces larger amounts of airborne pollen because of its overall size compared to the average size of the common ragweed.

COMPOSITAE

Iva axillaris Pursh   Povertyweed, small-flowered marsh elder.

Description. Grains tricolporate, rarely tetracolporate, polar area index 0.7–0.8; in polar view spheroidal, 21.0–23.5 μm, av 22.5 μm in diam; in meridional view slightly oblate, 21.0–22.0 μm wide × 20.5–21.0 μm long; sculpturing echinate; ektexine and endexine about 3.5 μm thick; structure tectate.

SEM: Colpi 4.0–4.5 μm long, the enclosed germ pore about 2 μm long, spines about 1 μm long, about 3 μm apart, tips sharply pointed.

Flowering. Early part of June, maximum period the latter part of June or early July, pollen shedding into the early part of August.

Native to. North America.

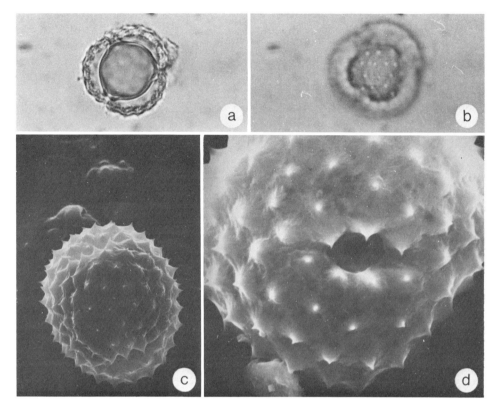

Fig. 70. *Iva axillaris.* a–b, light microscope, × 1000, a, polar view, equatorial section, b, surface; c–d, SEM, c, × 1800, d, × 4200.

Distribution. Common in the Prairie Provinces; a subspecies occurs in the interior of southern British Columbia (Bassett, Mulligan, and Frankton 1962).

British Columbia: Subspecies, *Iva axillaris* Pursh ssp. *robustior* (Hook.) Bassett, collected at three different locations in the interior of the province.

Alberta: Povertyweed including subspecies *robustior* widely scattered from the Peace River district to the USA, mainly in saline habitats.

Saskatchewan: South of the Saskatchewan River to the USA, widely common in alkaline habitats.

Manitoba: Common in alkaline habitats of southern Manitoba, especially west of the Red River.

Notes. The pollen-collecting devices established in the Prairie Provinces in 1962 near large stands of povertyweed did not accumulate any large pollen counts. Wodehouse (1971) stated that although individual plants shed only little pollen, they are frequently so numerous that they are an important factor in causing hay fever. Biological information on povertyweed in Canada has recently been prepared by Best (1975).

COMPOSITAE

Iva frutescens L.   Marsh elder, high-water shrub.

Description.   Grains tricolporate, rarely tetracolporate, polar area index 0.2; in polar view spheroidal, 19.5–22.5 μm, av 21.0 μm in diam; in meridional view oblate, 20.5–23.0 μm wide × 18.5–20.0 μm long; sculpturing echinate; ektexine and endexine about 2 μm thick; structure tectate.

SEM: Colpi 3.0–3.5 μm long, the enclosed germ pore about 2 μm long; spines about 1 μm long, about 3 μm apart, tips sharply pointed.

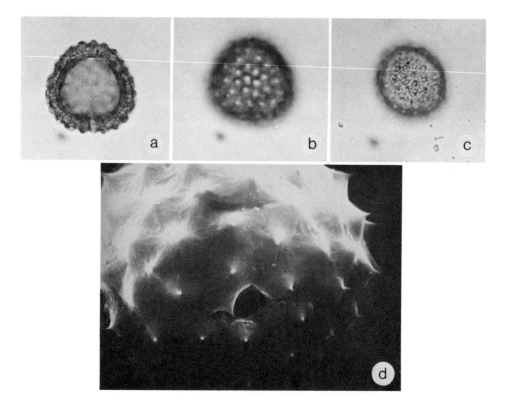

Fig. 71.   *Iva frutescens. a–c*, light microscope, × 1000, *a*, polar view, equatorial section, *b*, spines at high focus, *c*, surface at low focus; *d*, SEM, × 4200.

Flowering. The end of July, peaking the first 2 weeks of August, pollen shed until the end of August.

Native to. North America.

Distribution. A few locations in Nova Scotia only.

Notes. The pollen has not been found in large enough amounts to cause hay fever. Wodehouse (1971) stated that on account of its restricted range and general lack of abundance, this species is only a local and occasional cause of hay fever.

COMPOSITAE

Iva xanthifolia Nutt. False ragweed, burweed marsh elder, careless weed, prairie ragweed, marsh elder, horseweed.

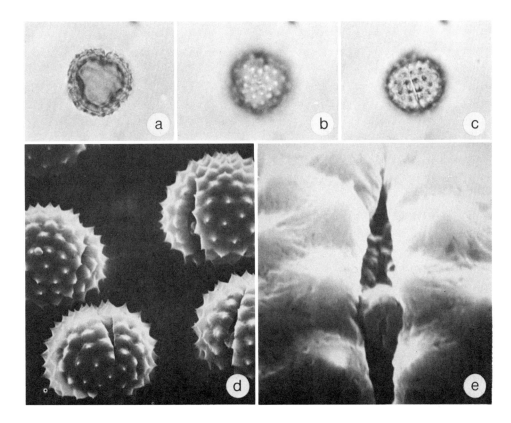

Fig. 72. *Iva xanthifolia. a–c*, light microscope, × 1000, *a*, polar view, equatorial section, *b*, surface, *c*, colpi; *d–e*, SEM, *d*, × 1680, *e*, × 8400.

Description. Grains tricolporate, polar area index 0.4–0.5; in polar view spheroidal, 18.0–20.0 μm, av 19.5 μm in diam; in meridional view oblate, rarely spheroidal, 19.5–21.0 μm wide × 16–18 μm long; sculpturing echinate; ektexine and endexine about 1.5 μm thick; structure tectate.

SEM: Colpi 10–12 μm long, the enclosed germ pore about 2 μm long; spines about 1 μm long, about 3 μm apart, tips sharply pointed.

Flowering. The first of August to the latter part of September, with a peak about the middle of August for 2 weeks.

Native to. North America.

Distribution. All provinces except Nova Scotia.

British Columbia: Small patches in the southern part of the province, especially in the Okanagan Valley, waste places and along roadsides near irrigated areas.

Alberta: Widely scattered patches along roadsides and waste places throughout the southern part of the province.

Saskatchewan: Common in the southern part of the province along roadsides and in waste places.

Manitoba: Common in the southern part of the province along roadsides and in waste places.

Ontario: Small patches occasionally along roadsides and waste places throughout the southern part of the province.

Quebec: Few small patches throughout the southern part of the province.

New Brunswick: Rare throughout the province.

Prince Edward Island: Collected in a few disturbed places, but rare.

Notes. Several of the pollen collecting devices were established near stands of false ragweed in Western Canada (Bassett 1964), but very little pollen was collected on the slides. This suggests that these plants do not shed large amounts of pollen at any one period of the flowering season. Wodehouse (1971) stated that false ragweed is one of the most important causes of hay fever throughout much of its range. Both its leaves and pollen are known to cause dermatitis. Further studies are needed to evaluate the potential of the pollen in causing hay fever.

COMPOSITAE

Solidago canadensis L.   Canada goldenrod, goldenrod.

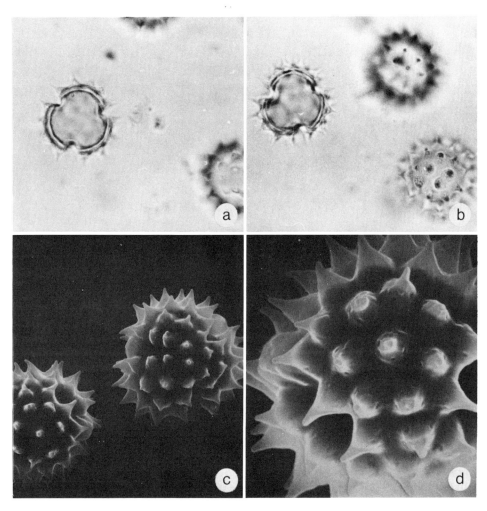

Fig. 73. *Solidago canadensis. a–b*, light microscope, × 1000, *a*, polar view, equatorial section, *b*, 3 grains in various positions and focus; *c–d*, SEM, *c*, × 1800, *d*, × 4570.

Description. Grains tricolporate, occasionally tetracolporate, polar area index 0.4; in polar view spheroidal, 23–28 μm, av 25 μm in diam; in meridional view spheroidal; sculpturing echinate; ektexine and endexine about 1.5 μm; structure tectate.

SEM: Colpi about 12 μm long, the enclosed germ pore about 2 μm long; spines 3.5–4.0 μm long, several distinct pores at the base of the spines, tips rounded and not pointed.

Flowering. Early July to the latter part of September.

Native to. North America.

Distribution. All provinces and very common in several areas.

Notes. Pollen grains of some insect-pollinated plants, such as the Canada goldenrod, can produce hay fever, but normally the heavy sticky pollen is carried by insects or drops to the ground close to the plants (Bassett and Frankton 1971). Aster pollen is similar to that of the goldenrod.

COMPOSITAE

Xanthium spinosum L.   Spiny cocklebur, California cocklebur.

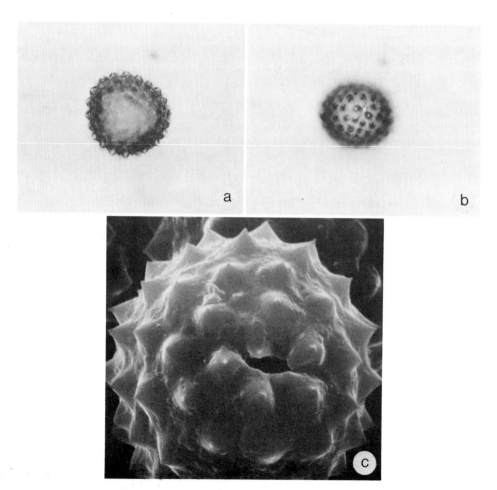

Fig. 74. *Xanthium spinosum. a–b*, light microscope, × 1000, *a*, polar view, equatorial section, *b*, surface and aperture; *c*, SEM, × 4000.

Description. Grains tricolporate, rarely tetracolporate, polar area index large to very large; in polar view spheroidal, 19.0–25.0 μm, av 22.0 μm in diam; in meridional view spheroidal or slightly oblate, 20.0–26.0 μm wide × 19.0–24.0 μm long; sculpturing microechinate; ektexine and endexine about 1.5 μm thick; structure tectate.

SEM: Colpi 2.5–3.5 μm long, the enclosed germ pore not clearly defined; spines approximately 60 on half of the pollen surface, less than 1 μm long, about 3 μm apart, tips not sharply pointed.

Flowering. Early July to early August, with a peak the last 2 weeks of July.

Native to. Europe.

Distribution. Only known from a few locations in southern Ontario.

Notes. Spiny cocklebur is not weedy or abundant enough in Canada to shed sufficient pollen to cause hay fever.

COMPOSITAE

Xanthium strumarium L. Cocklebur, clothur.

Description. Grains tricolporate, rarely tetracolporate, polar area index 0.8; in polar view spheroidal, 24.5–26.0 μm, av 25.5 μm in diam; in meridional view slightly oblate, 24.0–27.0 μm wide × 23.0–26.0 μm long; sculpturing microechinate; ektexine and endexine about 1.5 μm thick; structure tectate.

SEM: Colpi about 3 μm long, the enclosed germ pore about 1.5 μm long; spines approximately 90 on half of the pollen surface, about 0.5 μm long, about 2 μm apart, tips not sharply pointed.

Flowering. Around July 10 to early August, peaking the last 2 weeks in July.

Native to. North America.

Distribution. Widely scattered at the lower latitudes, but not generally common.

Notes. Because very few pollen grains of the cockleburs have been caught on exposed slides in different parts of the country, it is doubtful if they play much of a role in causing hay fever. The pollen is known to cross-react with that of ragweed (Wodehouse 1971), but most species shed too little to be more than a minor contributing factor in hay fever.

(Fig. 75 overleaf)

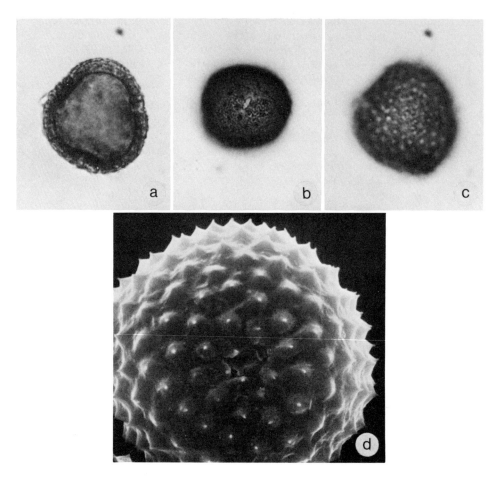

Fig. 75. *Xanthium strumarium. a-c,* light microscope, × 1000, *a,* polar view, equatorial section, *b,* surface and aperture, *c,* surface of polar end at high focus; *d,* SEM, × 4000.

Pollen Key to the Species of *Artemisia* (Wormwoods or Sages)

A. Grains averaging less than 20 μm in diam (polar view); enclosed pore under the SEM about 4.5 μm long..***Artemisia campestris***

A. Grains averaging more than 20 μm in diam (polar view); enclosed pore under the SEM less than 4.5 μm long (except for *A. frigida* and *A. vulgaris*)

    B. Spines under the SEM about 3 μm apart .... ***A. biennis***

    B. Spines under the SEM about 1 μm apart

C. Colpi under the SEM less than 5 μm long .... ***A. frigida***

C. Colpi under the SEM more than 5 μm long

    D. Spines under the SEM less than 0.5 μm long .... ***A. ludoviciana***
                                                                                                      A. vulgaris

    D. Spines under the SEM more than 0.5 μm long .... ***A. tridentata***

COMPOSITAE

Artemisia biennis Willd.   Biennial wormwood, sagewort.

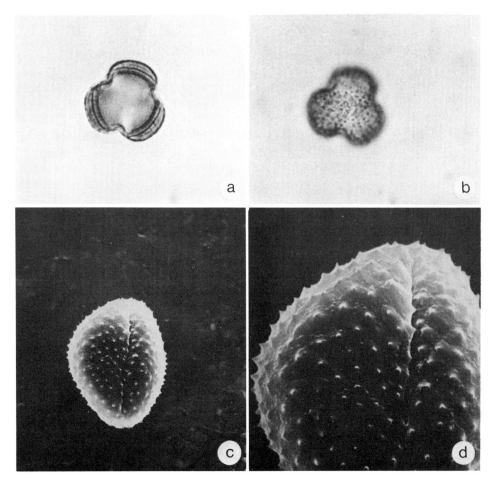

Fig. 76. *Artemisia biennis*. *a–b*, light microscope, × 1000, *a*, polar view, equatorial section, *b*, polar surface; *c–d*, SEM, *c*, × 1650, *d*, × 4000.

Description. Grains tricolporate, polar area index 0.3; in polar view spheroidal, 20.0–21.5 µm, av 21.0 µm in diam; in meridional view spheroidal to slightly oblate or occasionally prolate, 16.5–20.0 µm wide × 20.5–22.5 µm long; sculpturing microechinate; ektexine and endexine about 3 µm thick; structure tectate.

SEM: Colpi about 15 µm long, the enclosed germ pore 1.0–1.5 µm long; spines 0.5–1.0 µm long, about 3 µm apart, tips not sharply pointed.

Flowering. About the first of August to the end of September, peaking the last 2 weeks of August and the first week of September.

Native to. North America.

Distribution. Widely scattered and occasionally common in open clearings, roadsides, and waste places from British Columbia to Quebec; not as common or widely scattered in the Maritime Provinces.

Notes. Large amounts of pollen are shed in the late summer and fall seasons and this causes hay fever.

COMPOSITAE

Artemisia campestris L.   Field sagewort, perennial sagewort.

Description. Grains tricolporate, occasionally tetracolporate, polar area index 0.2–0.3; in polar view spheroidal, 16.5–19.5 µm, av 18.0 µm in diam; in meridional view oblate and occasionally prolate, 17.5–20.0 µm wide × 14.5–17.5 µm long; sculpturing microechinate; ektexine and endexine about 2 µm thick, structure tectate.

SEM: Colpi about 16 µm long, the enclosed germ pore about 4.5 µm long; spines less than 0.5 µm long, about 1 µm apart, tips not sharply pointed.

Flowering. Latter part of July to September, peaking in early August.

Native to. Europe and Asia.

Distribution. Widespread in every province and in the Yukon.

Notes. Pollen shed in large amounts can cause hay fever.

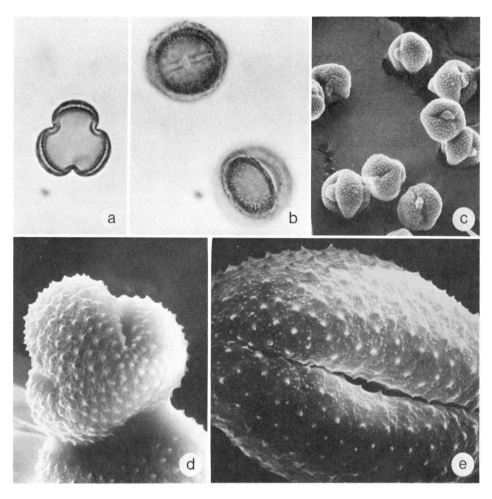

Fig. 77. *Artemisia campestris. a-b*, light microscope, × 1000, *a*, polar view, equatorial section, *b*, 2 grains in different positions and focus; *c-e*, SEM, *c*, × 800, *d*, × 3850, *e*, × 4000.

COMPOSITAE

Artemisia frigida Willd.   Pasture sage, prairie sage, carpet sage, fringed sagebrush, Arctic sagebrush, estafiata.

Description.   Grains tricolporate, rarely tetracolporate, polar area index 0.3; in polar view spheroidal, 19.5–21.5 μm, av 20.5 μm in diam; in meridional view oblate, 21.0–22.5 μm wide × 15.0–18.5 μm long; sculpturing microechinate; ektexine and endexine about 2.5 μm thick; structure tectate.

165

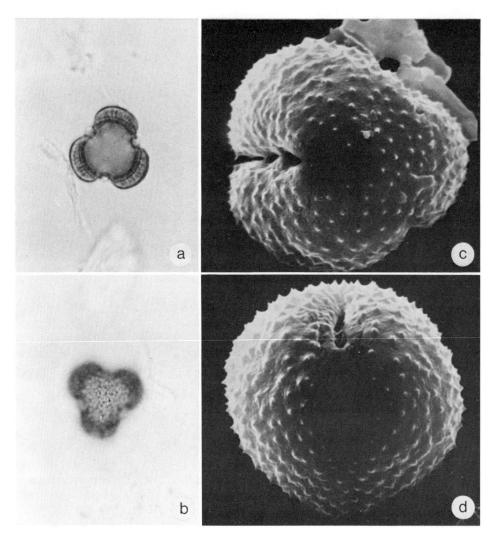

Fig. 78. *Artemisia frigida.* a–b, light microscope, × 1000, a, polar view, equatorial section, b, polar surface; c–d, SEM, × 4000.

SEM: Colpi 2.4–4.5 μm long, the enclosed germ pore about 1 μm long; spines less than 0.5 μm long, about 1 μm apart, tips not sharply pointed.

Flowering. First of August to September, peaking during the last 2 weeks of August and the first week of September.

Native to. North America.

Distribution. Yukon and from British Columbia to New Brunswick; widely scattered in the interior of southern British Columbia in the drier areas and in the Prairie Provinces; rare in Ontario, Quebec, and New Brunswick.

Notes. According to Wodehouse (1971) the pollen causes much hay fever in areas where it is abundant.

COMPOSITAE

Artemisia ludoviciana Nutt. Prairie sage, green sage, dark-leaved mugwort, wormwood, white sagebrush.

Description. Grains tricolporate, occasionally tetracolporate, polar area index 0.2–0.3; in polar view spheroidal, 21.5–25.5 μm, av 23.5 μm in diam; in meridional view oblate and prolate, 20.0–23.0 wide × 22.0–26.5 μm long; sculpturing microechinate; ektexine and endexine about 2 μm thick; structure tectate.

SEM: Colpi about 18 μm long, the enclosed germ about 3 μm long; spines less than 0.5 μm long, about 1 μm apart, tips rounded.

Flowering. First of August to the end of September, peaking the last week of August and the first week of September.

Native to. North America.

Distribution. More common in the Prairie Provinces than the interior of British Columbia; rare in Ontario and Quebec.

Notes. Where common the pollen can cause hay fever.

(Fig. 79 overleaf)

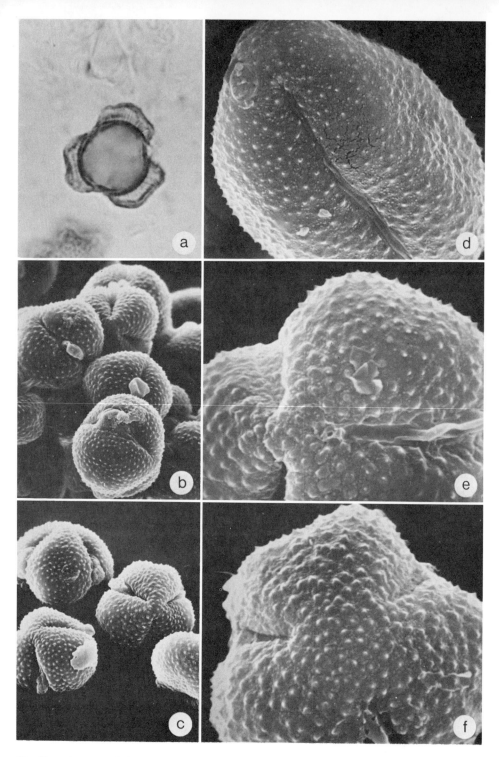

Fig. 79. *Artemisia ludoviciana. a*, polar view, equatorial section, light microscope, × 1000; *b–f*, SEM, *b–c*, × 1550, *d–f*, × 3900.

COMPOSITAE

Artemisia tridentata Nutt.   Big sagebrush, basin sagebrush, mountain sagebrush, black sage.

Description. Grains tricolporate, polar area index 0.2–0.3; in polar view spheroidal, 21–24 μm, av 23 μm in diam; in meridional view oblate or occasionally prolate, 19.0–22.5 μm wide × 20–24 μm long; sculpturing microechinate; ektexine and endexine about 2 μm thick; structure tectate.

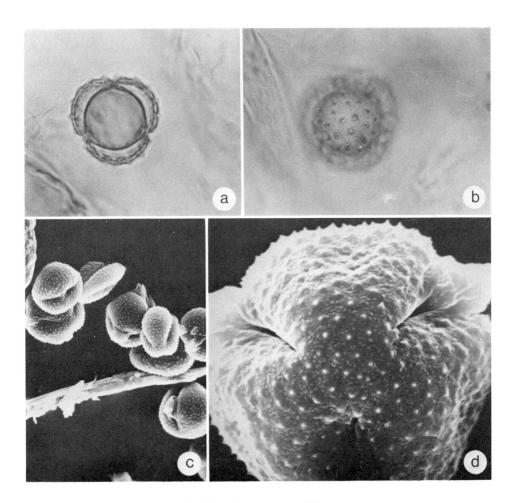

Fig. 80. *Artemisia tridentata. a–b*, light microscope, × 1000, *a*, polar view, equatorial section, *b*, surface; *c–d*, *SEM*, *c*, × 800, *d*, × 4100.

SEM: Colpi about 15 μm long, the enclosed germ pore about 2 μm long; spines 1.0–1.5 μm long, about 1 μm apart, tips not sharply pointed.

Flowering. First week of August to mid-October, peaking the last week of August and the first 2 weeks of September.

Native to. North America.

Distribution. British Columbia and Alberta, common in the interior of southern British Columbia in the dry areas; found in southern Alberta at only two locations.

Notes. According to Wodehouse (1971) it is the most abundant and important of all the sagebrushes as a cause of hay fever and in many places greatly outranks the ragweeds. Although Wodehouse was primarily discussing the importance of this species in the western United States, there is no doubt that in the interior of British Columbia enough pollen is shed to cause hay fever.

COMPOSITAE

Artemisia vulgaris L.   Mugwort, common wormwood, common sagewort.

Description. Grains tricolporate, polar area index 0.2; in polar view spheroidal, 20.0–23.0 μm, av 21.5 μm in diam; in meridional view spheroidal to slightly oblate or prolate, 18.5–24.0 μm wide × 15.5–19.0 μm long; sculpturing microechinate; ektexine and endexine about 2 μm thick, structure tectate.

SEM: Colpi about 10 μm long, the enclosed pore about 5 μm long; spines less than 0.5 μm long, about 1 μm apart, tips not sharply pointed.

Flowering. Around August to the end of September, peaking the last two weeks of August and the first week of September.

Native to. Europe and Asia.

Distribution. Scattered from British Columbia to Newfoundland.

Notes. Wodehouse (1971) stated that mugwort sheds fairly large amounts of pollen, but the quantity is seldom abundant enough to be of much importance in causing hay fever.

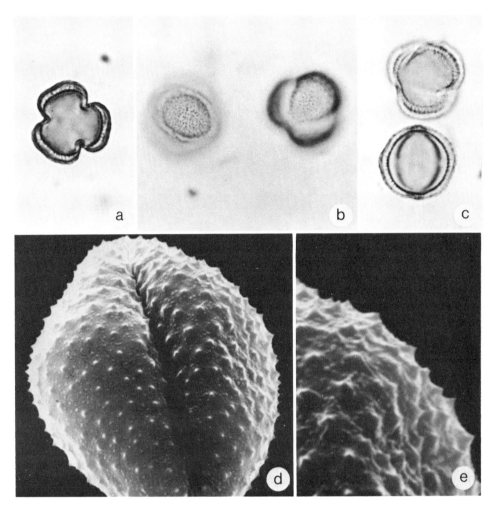

Fig. 81. *Artemisia vulgaris.* a–c, light microscope, × 1000, a, polar view, equatorial section, b, 2 grains in different positions and focus, c, equatorial view and section; d–e, SEM, d, × 4000, e, × 8000.

CRUCIFERAE

Cardamine pratensis L.   Bitter cress, cuckooflower, lady's-smock.

Description. Grains tricolpate, rarely tetracolpate; prolate, rarely circular or ovoid, 26–36 μm wide × 36–44 μm long, av 33 × 38 μm; polar area index variable and very inconsistent because of prolate grains with longer furrows and less width than the oblate grains; sculpturing reticulate; ektexine 2.0 μm, endexine 0.5 μm, wall 2.5 μm; structure tectate, baculate.

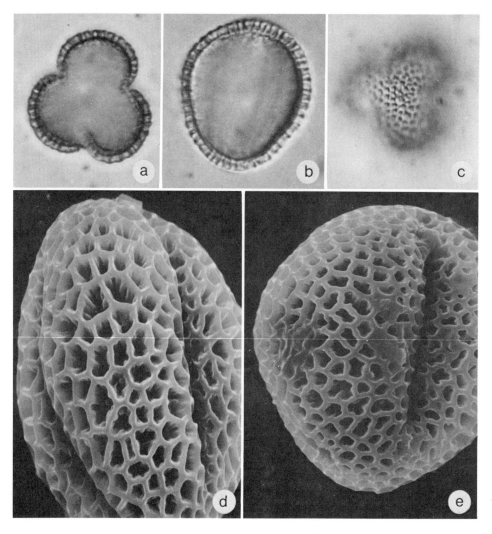

Fig. 82. *Cardamine pratensis.* a–c, light microscope (interference contrast), × 1000, a, polar view, equatorial section, b, equatorial view and section, c, polar end; d–e, SEM, × 2000.

SEM: Sculpturing reticulate, supported by micropilae, some two-rowed nature; luminae irregular, 0.4–1.0 μm in diam; muri rounded at the top.

Flowering. July.

Native to. North America.

Distribution. British Columbia to Newfoundland.

Notes. *Cardamine pratensis* var. *angustifolia* Hook. occurs from Alaska to Greenland. Very rarely is any pollen of the mustard family, which is not wind-pollinated, caught on exposed slides. The pollen is not known to cause hay fever.

CYPERACEAE

Carex aquatilis Wahlenb.   Sedge, aquatic sedge.

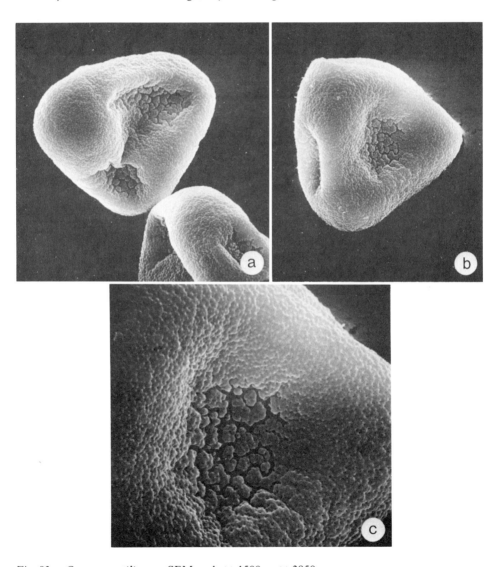

Fig. 83. *Carex aquatilis. a–c*, SEM, *a–b*, × 1500, *c*, × 3850.

Description. Grains inaperturate; irregular ovoid to pear-shaped with 3–6 lacunae, 35–47 μm, av 41 μm in diam; lacunae intruding, 15 μm in diam, circular to bowl-shaped, covered with a granular membrane of unevenly verrucate particles; sculpturing undetermined; ektexine and endexine of equal thickness, wall 1 μm thick; structure tectate with small baculate particles.

SEM: Minute perforations in the undulating exine surface that possesses scabrate elements; covering over the pore has verrucate particles of unequal sizes with spaces in between showing the smooth endexine surface.

Flowering. June and July.

Native to. North America.

Distribution. Widespread from British Columbia to Newfoundland.

Notes. The genus *Carex* comprises a large number of species that are widely distributed across Canada. Although many species possess pollen grains that are pear-shaped, there are grains that are irregular. Some species shed a lot of pollen, which has been occasionally suspected of causing hay fever (Wodehouse 1971). *Cyperus esculentus* L., yellow nut sedge, which is widely scattered in southern Ontario and Quebec, has also been suspected of causing hay fever.

ELAEAGNACEAE

Shepherdia argentea Nutt.   Silvery buffaloberry, buffaloberry.

Description. Grains tricolporate, resembling two, three-sided pyramids base to base when optically sectioned through the grain equator, 26–33 μm, av 30 μm in diam; polar area index 0.1–0.14; pores not clearly defined; sculpturing rugulate; endexine thicker at the equator, wall 2.5 μm thick; structure tectate.

SEM: Sculpturing rugulate with long colpi extending into the polar region.

Flowering. Mid-May.

Native to. North America.

Distribution. Widespread from Alberta to Manitoba.

Notes. A few pollen grains have been caught on exposed slides (Kennedy 1953), but they are now known to cause hay fever. The pollen of *Elaeagnus commutata* Bernh., silverberry, is similar to that of *Shepherdia argentea* except that the furrows are shorter and the sculpturing is rather smooth.

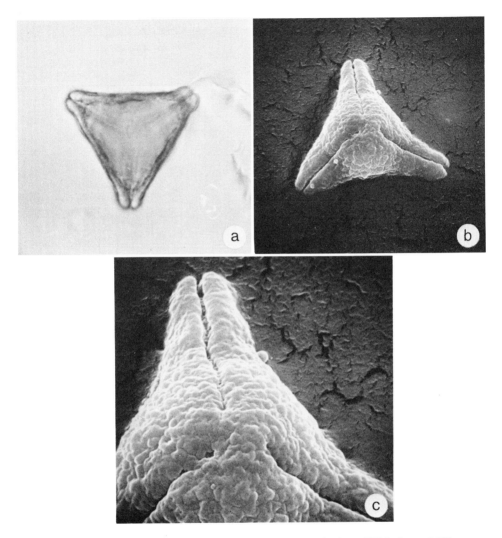

Fig. 84. *Shepherdia argentea*. a, light microscope, × 1000; b–c, SEM, b, × 1470, c, × 3650.

ELAEAGNACEAE

Shepherdia canadensis (L.) Nutt.   Russet, buffaloberry, soapberry.

Description.   Grains tricolporate, prolate, 17–21 μm wide × 28–36 μm long, av 19 × 32 μm; polar area index 0.2; pores circular, slightly annulate, 1 μm in diam, annulus 0.5 μm in diam; sculpturing rugulate; wall 1–5 μm thick; structure tectate.

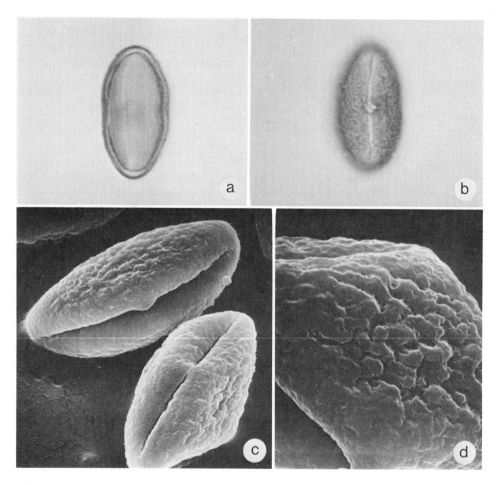

Fig. 85. *Shepherdia canadensis. a–b*, light microscope, × 1000, *a*, equatorial view and section, *b*, colpi and pore; *c–d*, SEM, *c*, × 1470, *d*, × 3650.

SEM: Sculpturing rugulate with minute perforations in the exine surface; pores beaklike constrictions in the center of the colpi.

Flowering. May.

Native to. North America.

Distribution. Widespread from the Yukon to Newfoundland.

Notes. Only a small amount of airborne pollen has been caught on exposed slides. It is not known to cause hay fever. The pollen of *Shepherdia* and *Elaeagnus* spp. often occurs in pollen profiles of the Pleistocene epoch.

FAGACEAE

Castanea dentata (Marsh.) Borkh.   Chestnut, sweet chestnut.

Description. Grains tricolporate; polar area index 0.33–0.46; in polar view circular, 8–11 μm wide; in meridional view prolate, 11–16 μm long; pores elliptical, 3 μm long × 1 μm wide, slightly constricted in the middle, split by the colpi and encircled by an elliptical-annular thickening slightly larger than the pore opening; colpi smooth margins; ektexine and endexine thickness difficult to determine, wall thickness 1 μm; sculpturing and structure difficult to determine, some grains with a mottled (reticulate) appearance.

SEM: Sculpturing rugulate, mainly limited to the area of the longitudinal intercolpium, ends mainly smooth; pores evident on grains with open colpi.

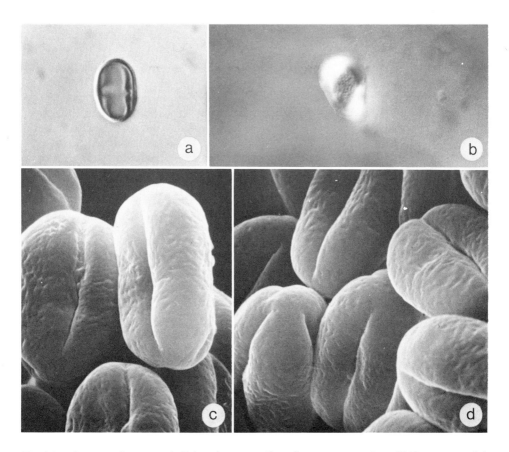

Fig. 86.   *Castanea dentata. a–b*, light microscope (interference contrast), × 1000, *a*, equatorial view and section, *b*, surface; *c–d*, SEM, × 3850.

Flowering. June and July.

Native to. Eastern North America.

Distribution. Southern Ontario only.

Notes. This species is sometimes confused with the European sweet chestnut, *Castanea sativa* Mill., and the horse chestnut, *Aesculus hippocastanum* L. *Castanea dentata* is not considered important in causing hay fever because it is not wind-pollinated.

FAGACEAE

Fagus grandifolia Ehrh. Beech, American beech, red beech.

Description. Grains tricolporate, occasionally dicolporate or tetracolporate; polar area index 0.2–0.35; in polar view circular, 29–44 μm, av 37 μm wide; in meridional view spheroidal or oblate, 29–39 μm, av 34 μm long; pores elliptical, 6–10 μm long, 3–4 μm wide, slightly constricted in the center, split by the colpi and encircled by an elliptical-annular thickening slightly larger than the pore opening; colpi granular with thickened margins; sculpturing of slightly raised verrulose-scabrate processes; ektexine and endexine of equal thickness; structure intectate with densely pilate wall, 1–2 μm thick.

SEM: Sculpturing microrugulate or microscabrate, irregularly positioned densely spaced aciculate trichomes.

Flowering. Latter part of April and early May.

Native to. Central and Eastern North America.

Distribution. Southern Ontario to Nova Scotia.

Notes. The pollen is considered as only a minor contributory factor in hay fever.

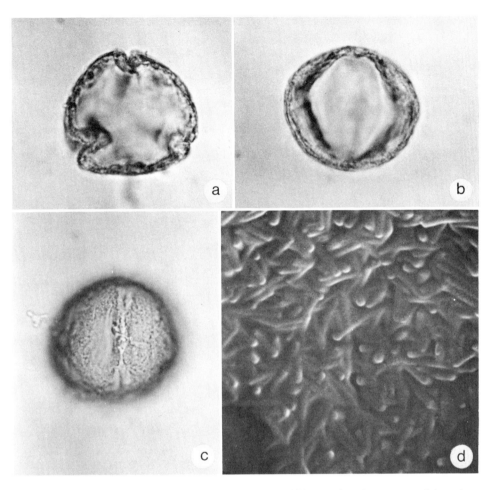

Fig. 87. *Fagus grandifolia*. a–c, light microscope, × 1000, a, polar view, equatorial section, b, equatorial view and section, c, colpi; d, SEM, × 9100.

Pollen Key to Eight Species of *Quercus* (Oaks)

A. Grains tricolporate

    B. Grains 20–24 μm, av 22 μm in diam (meridional view) .... **Quercus rubra**

    B. Grains 25–33 μm, av 28–29 μm in diam (meridional view)

        C. Ektexine and endexine of equal thickness .... **Q. palustris**

        C. Ektexine one-half the thickness of the endexine .... **Q. velutina**

A. Grains tricolpate

    D. Polar area index 0.18–0.42

        E. Wall 1 μm thick .... **Q. bicolor**

        E. Wall 2 μm thick .... **Q. muehlenbergii**

    D. Polar area index 0.3–0.6

        F. Sculpturing under the SEM, microechinate glomerate particles .... **Q. macrocarpa**

        F. Sculpturing under the SEM, not of microechinate glomerate particles

            G. Sculpturing under the SEM, evenly spaced verrucate particles .. **Q. alba**

            G. Sculpturing under the SEM, irregular spaced scabrate tubercles .... **Q. garryana**

FAGACEAE

Quercus alba L.   White oak, stove oak.

Description.   Grains tricolpate, polar area index 0.30–0.44; in polar view circular, 24–30 μm wide; in meridional view spheroidal, oblate, or prolate, 23–27 μm, av 25 μm long; colpi open, intruding, margins thickened; sculpturing coarsely irregular shaped verrucate particles sometimes echinate or scabrate; ektexine and endexine of equal thickness, wall 1.5–2.0 μm thick; structure tectate.

    SEM: Sculpturing evenly spaced verrucate particles 0.1–0.8 μm in diam with microperforations in the exine between the verrucae.

Flowering.   From the 15th to the end of May.

Native to.   Eastern North America.

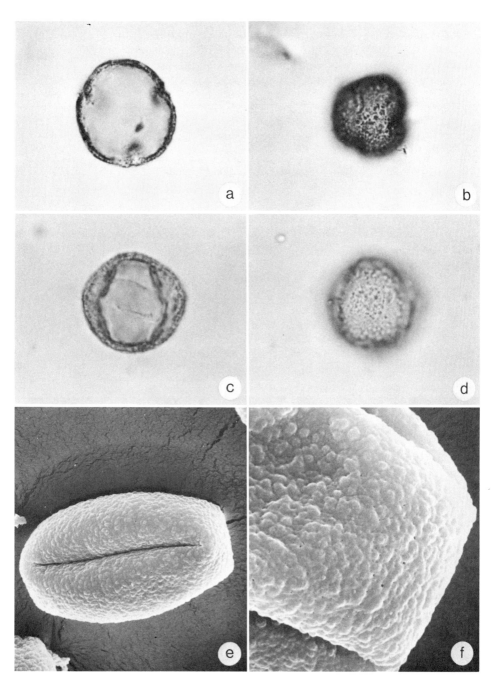

Fig. 88. *Quercus alba.* a–d, light microscope, × 1000, a, polar view, equatorial section, b, polar end, c, equatorial view and section, d, equatorial surface; e–f, SEM, e, × 1470, f, × 3700.

Distribution.   Southeastern Ontario to southern Quebec.

Notes.   All species of oak have pollen with the same allergenic qualities so that any large amounts in the air can cause hay fever.

FAGACEAE

Quercus bicolor Willd.   Swamp white oak, blue oak.

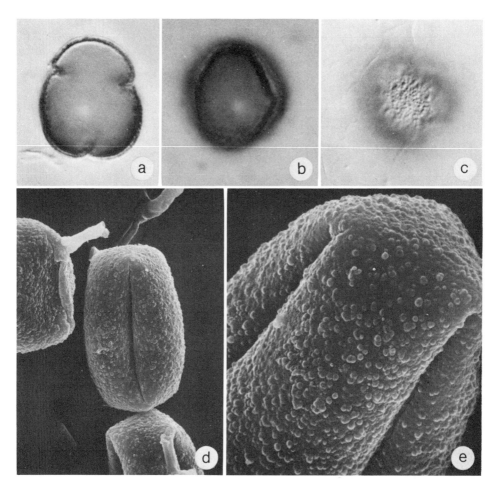

Fig. 89. *Quercus bicolor. a–c*, light microscope (interference contrast), × 1000, *a*, polar view, equatorial section, *b*, colpi, *c*, polar surface; *d–e*, SEM, *d*, × 1500, *e*, × 3700.

Description. Grains tricolpate; polar area index 0.19-0.36; in polar view circular, 23-29 μm, av 26.5 μm wide; in meridional view prolate, 19-26 μm, av 23 μm long; colpi opening quite wide, intruding, margins not conspicuously thickened; sculpturing irregular verrucate particles; ektexine and endexine of equal thickness, wall 1.0 μm thick; structure tectate.

SEM: Sculpturing verrucate-tuberculate particles 0.1-0.8 μm in diam; particles irregularly arranged on the grain surface.

Flowering. June.

Native to. North America.

Distribution. Southwestern Ontario to southwestern Quebec.

Notes. In large amounts the pollen from this species can cause hay fever.

FAGACEAE

Quercus garryana Dougl.   Garry oak, Oregon white oak, western white oak, Pacific white oak.

Description. Grains tricolpate, often tetracolpate; polar area index 0.43-0.45; in polar view circular, 30-36 μm, av 33 μm wide; in meridional view oblate to prolate, 23-32 μm, av 28 μm long; colpi open, intruding, margins slightly thickened; sculpturing coarsely irregular shaped, verrucate particles; ektexine and endexine indistinct, wall 2.5 μm thick; structure tectate.

SEM: Sculpturing irregularly spaced scabrate tubercles 0.3-0.8 μm in diam, arising from a verrucate-rugulate exine surface.

Flowering. June and July.

Native to. Western North America.

Distribution. The southwestern coastline of British Columbia.

Notes. Large quantities of pollen can cause hay fever.

(Fig. 90 overleaf)

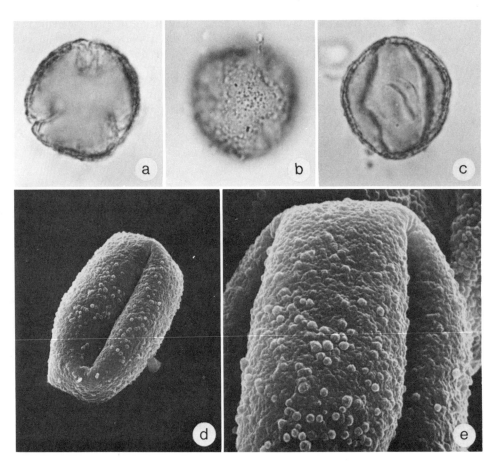

Fig. 90. *Quercus garryana*. a–c, light microscope, × 1000, a, polar view, equatorial section, b, polar surface, c, equatorial view and section; d–e, SEM, d, × 1470, e, × 3570.

FAGACEAE

Quercus macrocarpa Michx.   Bur oak, scrub oak, blue oak, mossy cup oak.

Description.   Grains tricolpate; polar area index 0.4–0.6; in polar view circular, 25–34 μm, av 29.5 μm wide; in meridional view oblate to prolate, 18–28 μm, av 24 μm long; colpi open, intruding, margins slightly thickened; sculpturing coarse, irregular, verrucate particles; ektexine and endexine of equal thickness, wall 2 μm thick; structure tectate.

SEM: Sculpturing microechinate, glomerulate particles 0.1–0.8 μm in diam, densely packed on the grain surface.

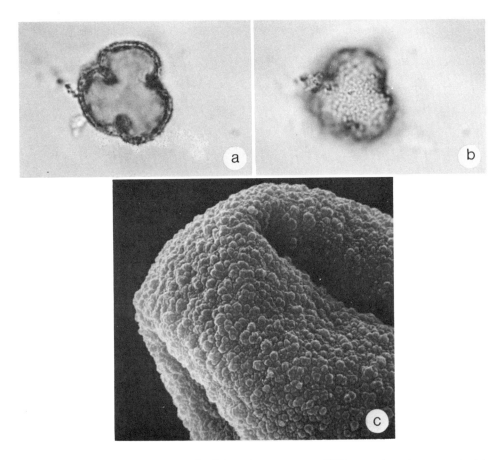

Fig. 91. *Quercus macrocarpa.* a–b, light microscope, × 1000, a, polar view, equatorial section, b, polar surface; c, SEM, × 3780.

Flowering. June.

Native to. North America.

Distribution. Saskatchewan to New Brunswick.

Notes. Large amounts of pollen from this species can cause hay fever.

FAGACEAE

Quercus muehlenbergii Engelm.   Chinquapin oak, yellow oak, chestnut oak.

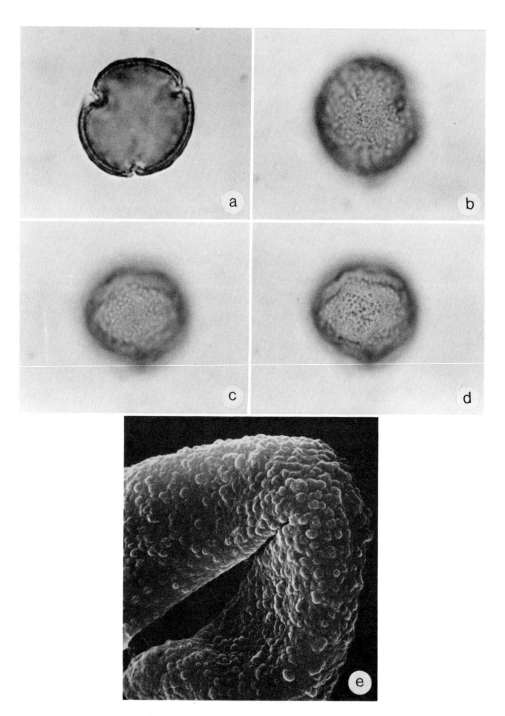

Fig. 92. *Quercus muehlenbergii*. *a–d*, light microscope, × 1000, *a*, polar view, equatorial section, *b*, polar surface, *c–d*, equatorial view in different focus; *e*, SEM, × 3700.

Description. Grains tricolpate; polar area index 0.18-0.42; in polar view circular, 19-30 μm, av 26 μm wide; in meridional view oblate to prolate, 19-27 μm, av 24 μm long; colpi open, intruding, margins thickened; sculpturing coarse verrucate irregular particles; ektexine thicker than the endexine, wall 2 μm thick; structure tectate.

SEM: Sculpturing similar to the grains of *Q. bicolor;* tuberculate particles 0.1-1.3 μm in diam.

Flowering. June.

Native to. North America.

Distribution. Southwestern Ontario.

Notes. Pollen from this species in large amounts can cause hay fever.

FAGACEAE

Quercus palustris Muenchh. Pin oak, water oak, swamp oak.

Description. Grains tricolporate; polar area index 0.25-0.50; in polar view mainly circular in shape, 26-34 μm, av 29 μm wide; in meridional view prolate, 25-33 μm, av 28 μm long; colpi intruding, granular with thickened margins; pores indistinct, somewhat masked by the intruding colpi (quite evident on expanded grains); sculpturing similar to the pollen of other *Quercus* spp.; ektexine and endexine of equal thickness, wall 1.5 μm thick; structure tectate.

SEM: Sculpturing and particles similar to the grains of *Q. macrocarpa.*

Flowering. April and May.

Native to. North America.

Distribution. Southwestern Ontario.

Notes. In large amounts the pollen shed from this species can cause hay fever.

(Fig. 93 overleaf)

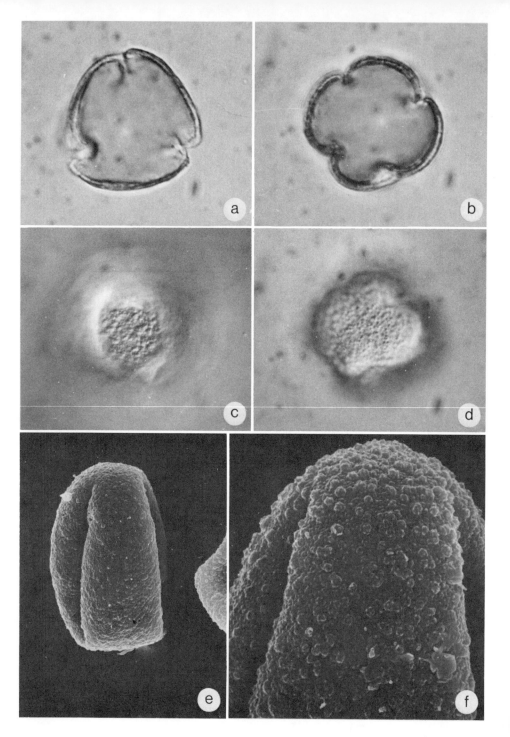

Fig. 93. *Quercus palustris*. a–d, light microscope, × 1000, a, polar view, equatorial section of tricolporate grain, b, equatorial section of tetracolporate grain, c–d, equatorial and polar surfaces under interference contrast; e–f, SEM, e, × 1470, f, × 3700.

FAGACEAE

Quercus rubra L.   Red oak, northern red oak, black oak.

Description. Grains tricolporate; polar area index 0.25–0.44; in polar view circular, 25–31 µm, av 27.5 µm wide; in meridional view oblate to subcircular, 20–24 µm, av 22 µm long; colpi closed, intruding, and with a distinct margin; pores not clearly defined, but show as constriction or a break in the colpi margin, partially masked by the closed colpi; sculpturing similar to pollen of other taxa in the genus; ektexine thinner than the endexine, wall 1.5–2.0 µm thick; structure tectate.

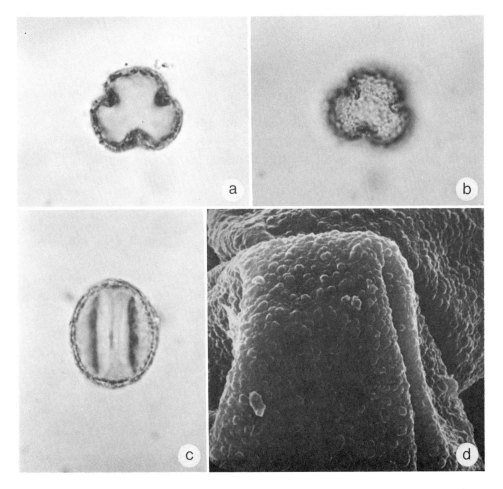

Fig. 94. *Quercus rubra.* a–c, light microscope, × 1000, a, polar view, equatorial section, b, view of polar end, c, equatorial view and section; d, SEM, × 3780.

SEM: Sculpturing similar to the grains of *Q. macrocarpa;* glomerulate particles 0.1–1.3 μm in diam.

Flowering. Late May and early June.

Native to. North America.

Distribution. Western Ontario to Nova Scotia.

Notes. In large amounts the pollen from this species can cause hay fever.

FAGACEAE

Quercus velutina Lam.   Black oak, yellow oak, yellow-barked oak, red oak.

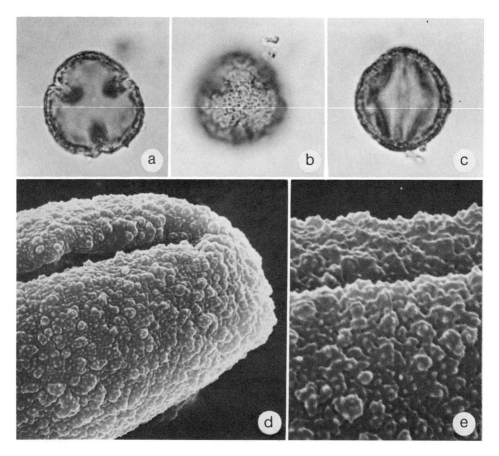

Fig. 95. *Quercus velutina. a–c,* light microscope, × 1000, *a,* polar view, equatorial section, *b,* view of polar end, *c,* equatorial view and section; *d–e,* SEM, *d,* × 3780, *e,* × 7500.

Description. Grains tricolporate; polar area index 0.25–0.35; in polar view circular, 22–26 μm, av 24 μm wide; in meridional view prolate to circular, 25–32 μm, av 29 μm long; colpi and pores similar to the pollen of *Q. rubra*; sculpturing coarse verrucate, elements mostly evenly spaced; ektexine half as thick as the endexine, wall 2 μm thick; structure tectate.

SEM: Sculpturing microechinate, glomerulate particles 0.3–1.0 μm in diam with microperforations in the exine between the glomerulate particles.

Flowering. Late May and early June.

Native to. North America.

Distribution. Southern Ontario.

Notes. When shed in large amounts, the pollen of this species can cause hay fever.

Pollen Key to Five Species of *Gramineae* (Grass Family)

Grains monoporate

A. Grains 85–125 μm, av 110 μm in diam .... ***Zea mays***

A. Grains 25–46 μm in diam

    B. Sculpturing under the SEM composed of evenly distributed microechinate particles .... ***Agropyron repens***

    B. Sculpturing under the SEM composed of evenly distributed irregular shaped, clusterlike glomerules topped with verrucate elements

        C. Operculum under the SEM occupies most of the pore area .... ***Dactylis glomerata***

        C. Operculum under the SEM occupies approximately half the pore area .... ***Phleum pratense***
                                                                                           Poa pratensis

GRAMINEAE

Agropyron repens (L.) Beauv.   Quack grass, couch grass, scotch grass, quitch grass, twitch grass.

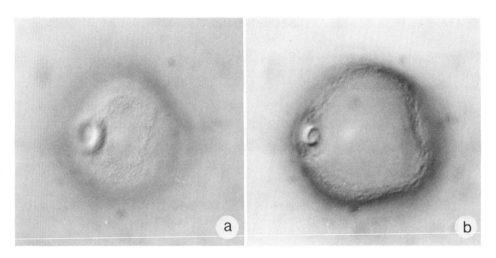

Fig. 96. *Agropyron repens. a–c,* light microscope (interference contrast), × 1000, *a,* annulus, *b,* annulus and operculum, *c,* surface; *d–e,* SEM, *d,* × 1750, *e,* × 4350.

Description.   Grains monoporate; spheroidal, 33–46 μm, av 40 μm in diam; pores circular in outline, 4 μm in diam, annulus 3.5–4.0 μm, not prominently elevated, operculum often destroyed by acetolysis; sculpturing microgranular particles; ektexine and endexine of equal thickness, wall 1.0 μm thick; structure tectate.

SEM: Sculpturing evenly distributed microechinate particles; operculum similar to that of *Phleum pratense.*

Flowering.   Mid-June to July and August, peaking near the end of June.

Native to.   Europe.

Distribution.   Widely scattered in all provinces and fairly abundant in some areas.

Notes.   The pollen is undoubtedly a contributing factor in hay fever.

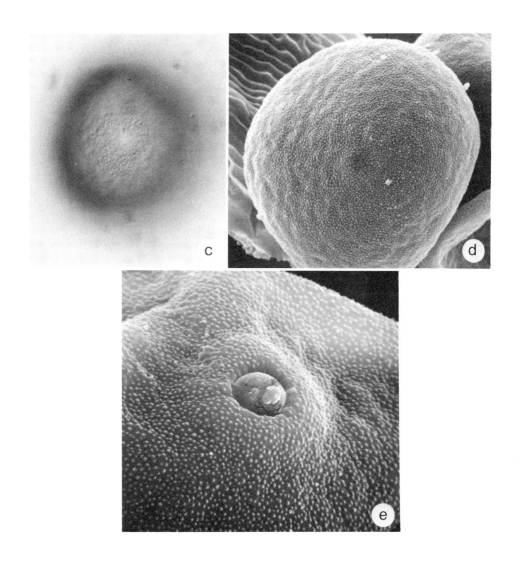

GRAMINEAE

Dactylis glomerata L.   Orchard grass, cocksfoot.

Description. Grains monoporate; in polar and meridional views spheroidal to ovoid, 28–38 μm, av 34 μm in diam; pores circular in outline, 3.5 μm in diam, annulus 2.5–3.0 μm in diam, operculum often destroyed by acetolysis; sculpturing microgranular particles in clusters; ektexine and endexine of equal thickness, wall 1.0 μm thick; structure tectate.

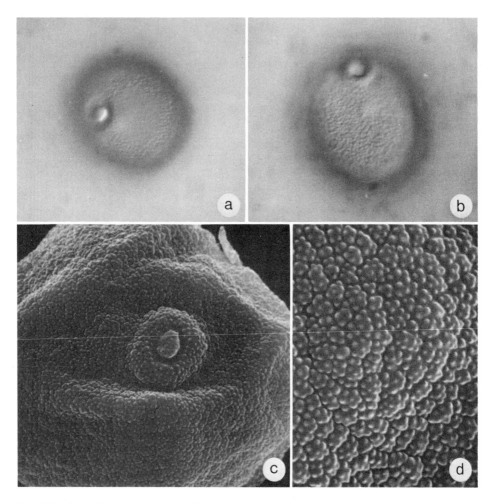

Fig. 97. *Dactylis glomerata. a–b*, light microscope (interference contrast), *a*, annulus, *b*, surface and annulus; *c–d*, SEM, *c*, × 3850, *d*, × 8050.

SEM: Glomerules evenly distributed, irregular shape, clusterlike topped with verrucate sculpturing elements; operculum, one large glomerule with verrucate sculpturing elements over most of the pore area; annulus, single verrucate particles.

Flowering. At Ottawa the mean date June 9 (Bassett et al. 1961) with peak about June 16 (Bassett 1956); extreme southern part of Ontario about a week earlier; Maritime Provinces (Bassett and Crompton 1969) and Western Canada about a week later than at Ottawa.

Native to. Europe.

Distribution. Cultivated in every province of Canada for producing hay; most common in southern British Columbia, Ontario, Quebec, and part of the Maritime Provinces.

Notes. Where this grass is abundant, the large amounts of pollen are considered one of the most important causes of summer hay fever.

GRAMINEAE

Phleum pratense L.   Timothy.

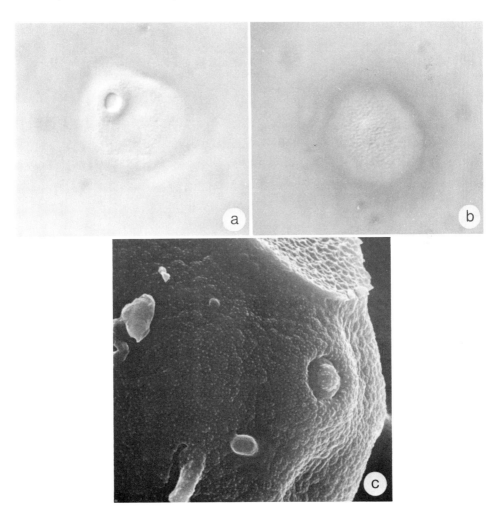

Fig. 98. *Phleum pratense. a–b*, light microscope, × 1000, *a*, annulus, *b*, surface; *c*, SEM, × 4000.

Description. Grains monoporate; in polar and meridional views spheroidal to ovoid, 28-38 µm, av 34 µm in diam; pores circular in outline, 4 µm in diam, annulus 2.5 µm, not prominent, operculum often destroyed by acetolysis; sculpturing microgranular particles; ektexine and endexine of equal thickness, wall 1.0 µm thick; structure tectate.

SEM: Sculpturing similar to that of *Dactylis glomerata,* the clusterlike glomerules not as clearly separated; annulus not prominent; operculum, circular microverrucate particle centered in the pore opening and occupying approximately half of the pore area.

Flowering. Ottawa June 26 with the peak period around the end of June and the first 2 weeks of July; in other parts flowering differs by 1 or 2 weeks depending on latitude.

Native to. Europe.

Distribution. Cultivated and widely naturalized in all provinces.

Notes. Where Timothy is common, it sheds large amounts of pollen and is considered one of the most important hay-fever grasses.

GRAMINEAE

Poa pratensis L. Kentucky blue grass, June grass, speargrass.

Description. Grains monoporate; in polar and meridional views spheroidal to ovoid, 25-40 µm, av 32 µm in diam; pores circular or rarely oval in outline, 3-5 µm in diam, annulus 3 µm in diam, operculum often destroyed by acetolysis; sculpturing microgranulate; ektexine and endexine mostly indistinct, ektexine appears thinner than the endexine, wall less than 1 µm thick; structure tectate.

SEM: Grains similar to those of *Phleum pratense.*

Flowering. Ottawa May 28 (Bassett et al. 1961), peaking the first 2 weeks of June; extreme southern Ontario the peak about 1 to 2 weeks earlier than at Ottawa; Maritime Provinces about 1 to 2 weeks later.

Native to. Europe and Asia.

Distribution. Common in all provinces.

Notes. The pollen of *Poa annua* L. is very similar to that of *P. pratensis,* which is considered important as a cause of hay fever. The pollen of *Agrostis*

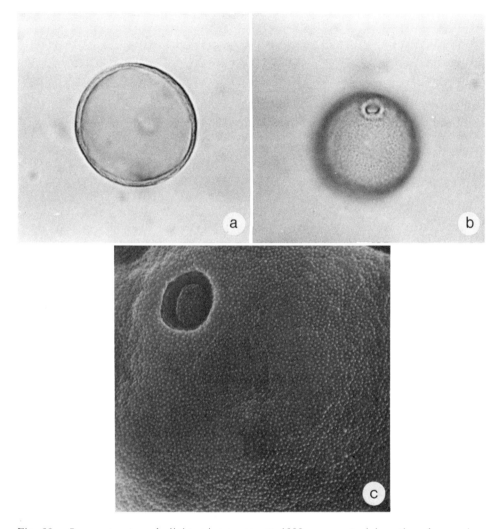

Fig. 99. *Poa pratensis:* a–b, light microscope, × 1000, a, equatorial section, b, annulus and surface; c, SEM, × 3850.

alba L., red top, and *Anthoxanthum odoratum* L., sweet vernal grass, are considered by Wodehouse (1971) as important in causing hay fever in the northeastern United States. Red top is widely scattered across Canada, but sweet vernal grass is not common except in some parts of southern Ontario and Quebec near the United States border.

GRAMINEAE

Zea mays L.   Corn, Indian corn, maize.

Description.   Grains monoporate; in polar and meridional views subspheroidal to ovoid, 85–125 μm, av 110 μm in diam; pores circular in outline, 6–7 μm in diam, annulus 6–7 μm in diam, operculum often destroyed by acetolysis; sculpturing microechinate; ektexine and endexine of equal thickness, wall 2 μm thick; structure tectate.

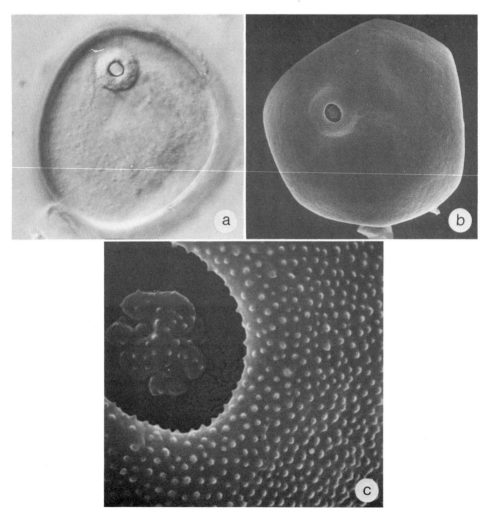

Fig. 100.   *Zea mays. a*, light microscope (interference contrast), × 600; *b–c*, SEM, *b*, × 770, *c*, × 8050.

SEM: Sculpturing microechinate with evenly distributed spine elements over the whole grain including the pore edge, exine surface otherwise smooth; operculum slightly recessed within the center of the pore and occupies approximately half of the irregularly shaped pore opening, microechinate spines.

Flowering.   July and August.

Native to.   North America.

Distribution.   Common in British Columbia, southern Ontario, Quebec, and the Maritime Provinces; some in southern Alberta and the other Prairie Provinces.

Notes.   The pollen of corn can cause hay fever but is considered too heavy to be transported very far from its source by the wind.

Pollen Key to the Bitternut, Pignut Hickory, and Shagbark, *Carya* spp.

Grains triporate

A. Grains 38–45 μm, av 41 μm in diam (meridional view) .... **Carya glabra**

A. Grains 32–44 μm, av 35–38 μm in diam (meridional view) .... **C. cordiformis**
C. ovata

JUGLANDACEAE

Carya cordiformis (Wang.) K. Koch   Bitternut hickory, swamp hickory.

Description.   Grains triporate; in polar view subisopolar, subtriangular, 39–46 μm, av 42 μm in diam; in meridional view elliptical, 33–38 μm, av 35 μm wide; pores circular, 3–4 μm in diam, evenly distributed near the grain equator, intruding; sculpturing scabrate; endexine slightly thicker than the ektexine except in the pore area where the endexine tapers slightly and the ektexine becomes tumescent, endexine on its interior surface appears torn and lacerated, wall 2 μm thick; structure tectate.

SEM: Sculpturing microechinate or scabrate elements evenly distributed on an otherwise smooth exine.

Flowering.   May and June.

Native to.   North America.

199

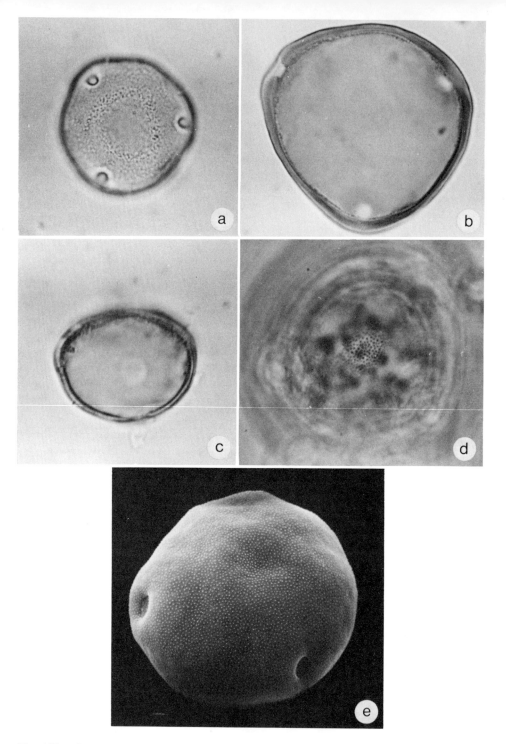

Fig. 101. *Carya* spp. *a–c*, light microscope, × 1000, *a*, polar view showing heteropolar pore arrangement, *b*, polar view, equatorial section, *c*, equatorial view and section, *d*, surface sculpture under phase contrast, × 1000; *e*, SEM, × 3850.

Distribution. Southern Ontario and southwestern Quebec.

Notes. Adams and Morton (1972) consider that pollen grains of *Carya* spp. occurring in Canada as seen under the SEM are inseparable. Whitehead (1963) has discussed pollen size frequency for several *Carya* species including *C. cordiformis, C. glabra,* and *C. ovata* in which he obtained approximately the same sizes as we did. Pollen shed in large amounts can cause hay fever.

JUGLANDACEAE

Carya glabra (Mill.) Sweet.   Pignut hickory, brown hickory, black hickory.

Description. Grains triporate, occasionally tetraporate; in polar view subisopolar, subtriangular, 45–55 µm, av 51 µm in diam; in meridional view elliptical, 38–45 µm, av 41 µm wide; pores circular, occasionally oblong, 3–5 µm in diam, evenly distributed near the grain equator, intruding; tetraporate grains, one or two smaller circular pores 1–2 µm in diam, unevenly spaced on the polar side of the grain; sculpturing scabrate; endexine twice as thick as the ektexine except near the pores where the ektexine becomes tumescent and the endexine narrower, interior of the endexine appears smooth, wall 2.5 µm thick; structure tectate.

SEM: Grains similar to those of *Carya cordiformis. See* Fig. 101.

Flowering. May and June.

Native to. North America.

Distribution. Southern Ontario.

Notes. Hybrids occur between *Carya* species where they overlap in distribution.

The pollen of pignut hickory can cause hay fever. See other notes under *C. cordiformis.*

JUGLANDACEAE

Carya ovata (Mill.) K. Koch   Shagbark hickory.

Description. Grains triporate; in polar view subisopolar, subtriangular, 40–52 µm, av 45 µm in diam; in meridional view elliptical, 32–44 µm, av 38 µm wide; pores circular to oblong, 3–5 µm in diam, evenly distributed near the grain equator, intruding; sculpturing scabrate; endexine slightly thicker than the ektexine except in the pore area where the endexine tapers slightly and the ektexine becomes tumescent, the interior surface of the endexine appears smooth or slightly pitted; structure tectate.

SEM: Grains similar to those of *Carya cordiformis*. See Fig. 101.

Flowering. May and June.

Native to. North America.

Distribution. Southern Ontario and southwestern Quebec.

Notes. Pollen from this species in large amounts can cause hay fever. See other notes under *C. cordiformis*.

JUGLANDACEAE

Juglans cinerea L.   Butternut, white walnut.

Description. Grains periporate, heteropolar; in polar view circular, 33–39 μm, av 36 μm in diam; in meridional view elliptical, 26–33 μm, av 29 μm wide; pores (7, 8-) 9 (-10, 12) on one side of the grain, unevenly distributed near the equator except for 1 or 2 pores at the polar end, 3 μm in diam, elliptical, annulate with thickenings up to 8 μm in diam (including pore); sculpturing microechinate; ektexine and endexine of equal thickness, wall 2 μm thick; structure tectate.

SEM: Sculpturing microechinate, particles evenly distributed over the grain surface; pores aspidate.

Flowering. June.

Native to. Eastern North America.

Distribution. Southern Ontario and Quebec to New Brunswick.

Notes. Based on pollen morphology, Whitehead (1963, 1965) discussed the taxonomy and phylogenetic trends in the family Juglandaceae including the genera *Juglans* and *Carya*.

Although it sheds large amounts of pollen at flowering time, the butternut is not considered as an important cause of hay fever. Vaughan and Black (1948) mentioned that patients frequently give positive reactions to its pollen, but treatment is not usually necessary.

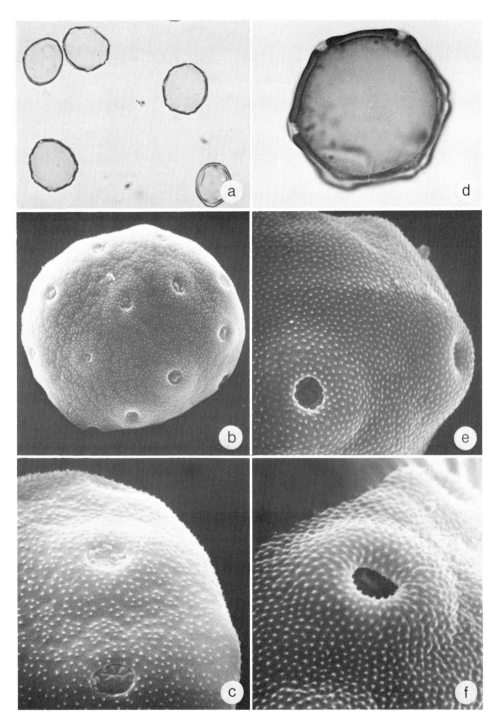

Fig. 102. *Juglans* spp. *a–c, J. cinerea, a*, light microscope, × 300; *b–c*, SEM, *b*, × 4000, *c*, × 5000. *d–f, J. nigra, d*, light microscope, × 1000, *e–f*, SEM, *e*, × 1600, *f*, × 4000.

JUGLANDACEAE

Juglans nigra L.   Black walnut, walnut, American walnut.

Description.   Grains periporate, heteropolar; in polar view circular, 36–42 μm, av 39 μm in diam; in meridional view elliptical, 29–35 μm, av 33 μm wide; pores (9, 10, 11–) 12 (–13, 14) on one side, unevenly distributed near the equator except for 1 or 2 at the polar end, 3–4 μm in diam, circular, with annular thickenings up to 10 μm in diam (including pores); sculpturing microechinate; ektexine and endexine of equal thickness, wall 2 μm thick; structure tectate.

SEM: Sculpturing microechinate, particles unevenly distributed over the grain surface; pores annulate with a granular membrane.

Flowering.   June.

Native to.   North America.

Distribution.   Southern Ontario.

Notes.   Although it sheds large amounts of airborne pollen, there is no evidence that this species causes hay fever, but occasionally it has been suspected of doing so (Wodehouse 1971). See other notes under *J. cinerea*.

JUNCACEAE

Luzula multiflora L.   Woodrush, field woodrush.

Description.   Grains united in tetrahedral tetrads with a triradiate scar separating the compartments, one large germ pore on the distal side of the grain in each compartment; grains 40–46 μm, av 43 μm in diam; sculpturing smooth; wall 1 μm thick except where the scar predominates; structure intectate without elevated elements on the scar part, and intectate with granular elements on the concave pores.

SEM: Sculpturing of two types: exine somewhat smooth with small scabrate or microechinate particles evenly distributed over the surface in areas not occupied by pores; pores with verrulose warts topped with microechinate spines evenly spaced to display a smooth endexine underneath.

Flowering.   June and early July.

Native to.   Europe and Asia.

Distribution.   British Columbia to Newfoundland.

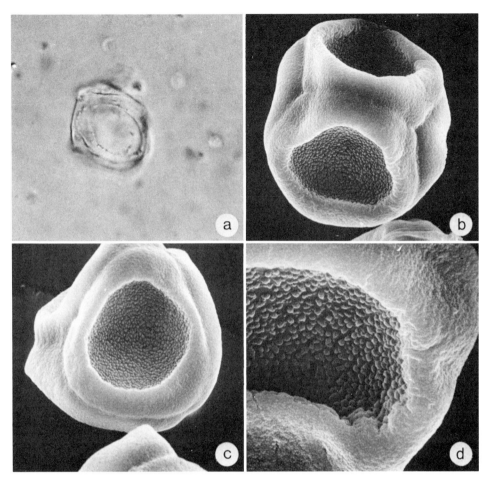

Fig. 103. *Luzula multiflora*. a, light microscope (interference contrast), × 1000; b–d, SEM, b–c, × 1575, d, × 3850.

Notes. Although woodrush is common in some areas and is wind-pollinated, its pollen is not known to cause hay fever.

JUNCAGINACEAE

Triglochin maritima L.   Seaside arrow-grass.

Description.   Grains inaperturate; spheroidal to irregular, 27–34 μm, av 30 μm in diam; easily distorted by acetolysis; sculpturing reticulate; complete wall 1 μm thick; structure intectate.

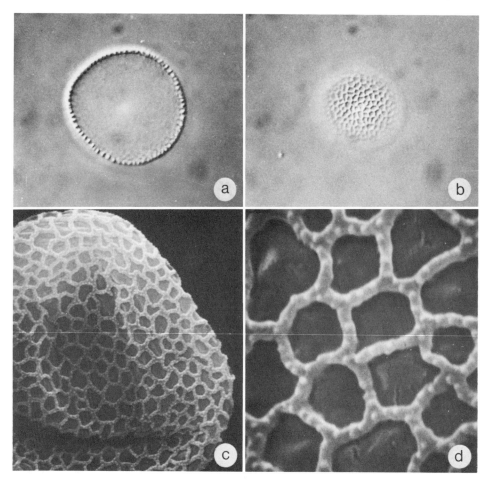

Fig. 104. *Triglochin maritima. a–b*, light microscope (interference contrast), × 1000, *a*, equatorial section, *b*, surface reticulation; *c–d*, SEM, *c*, × 4000, *d*, × 16 100.

SEM: Sculpturing irregular reticulate; tops of the muri essentially flat with evenly distributed spaced microgranules; reticulum on an otherwise smooth endexine.

Flowering. June and July.

Native to. North America.

Distribution. British Columbia to Newfoundland.

Notes. A few grains have been caught on exposed slides, but the pollen is not known to cause hay fever.

MORACEAE

Morus rubra L.   Red mulberry, mulberry, black mulberry.

Description.   Grains diporate, occasionally triporate; in polar view suboblate, rarely circular in shape, 19–22 μm, av 20 μm in diam; in meridional view, 16–19 μm, av 17 μm long; pores circular, 3 μm in diam with a granulate membrane; sculpturing scabrate or microechinate with processes evenly scattered over the grain; ektexine thicker than the endexine, wall less than 1 μm thick; structure tectate.

SEM: Sculpturing microechinate with processes evenly distributed over an otherwise smooth exine surface.

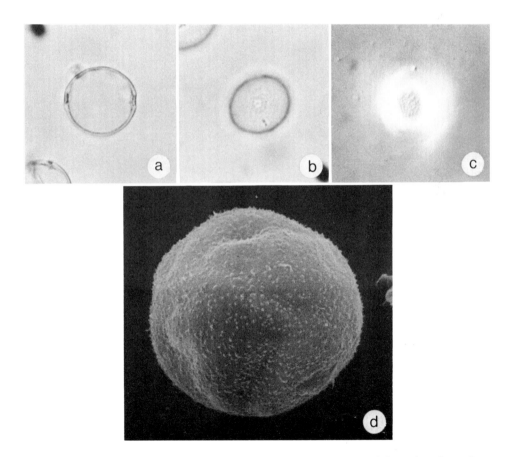

Fig. 105.   *Morus rubra.* a–c, light microscope, × 1000, a, equatorial section, b, surface showing pores, c, surface sculpturing under interference contrast; d, SEM, × 3650.

Flowering. May and the early part of June.

Native to. Eastern North America.

Distribution. Southern Ontario.

Notes. According to Wodehouse (1971), red mulberry is wind-pollinated and capable of causing very severe hay fever, but cases are relatively infrequent.

MYRICACEAE

Myrica gale L.   Sweet gale, meadow fern, piment royal.

Description. Grains triporate, rarely tetraporate; in polar view triporate grains circular to triangular, 25–35 μm, av 29 μm in diam; in meridional view elliptical, 20–28 μm, av 24 μm wide; pores evenly distributed around the equator of the grain, 2.5–3.5 μm in diam, circular, annulus (including pore) up to 10 μm in diam; sculpturing scabrate or microechinate; ektexine twice as thick as the endexine, endexine granular at the pore area, wall 2 μm thick; structure tectate.

SEM: Grains similar to those of *Betula papyrifera.*

Flowering. April to June.

Native to. North America.

Distribution. British Columbia to Newfoundland.

Notes. Sweet gale is known to shed large quantities of pollen that, according to Wodehouse (1971), possesses the physical characteristics of hay-fever pollen. Its role in hay fever is not fully understood and is confused by the fact that the pollen grains, when caught on pollen slides, are difficult to distinguish from those of the Betulaceae.

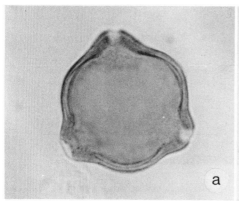
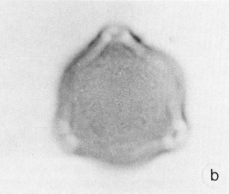

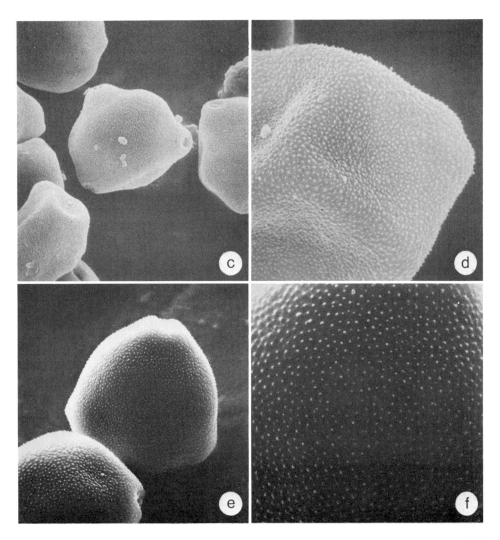

Fig. 106. *Myrica* spp. *a–d*, *M. gale*, *a–b*, light microscope, × 1000, *a*, polar view, equatorial section, *b*, surface; *c–d*, SEM, *c*, × 1400, *d*, × 3500. *e–f*, *M. pensylvanica*, SEM, *e*, × 1600, *f*, × 4000.

MYRICACEAE

Myrica pensylvanica Loisel.   Bayberry, candleberry.

Description. Similar to *M. gale* except for size of grain; in polar view grains circular to triangular, 22–28 μm, av 25 μm in diam; in meridional view elliptical, 19–24 μm, av 21 μm wide; sculpturing coarser scabrate than in *M. gale*.

SEM: Grains similar to those of *Betula papyrifera*.

Flowering. May to July.

Native to. Eastern North America.

Distribution. Ontario to Newfoundland.

Notes. At peak flowering when large amounts of pollen are shed, bayberry may cause hay fever.

MYRICACEAE

Comptonia peregrina (L.) Coult.   Sweet fern.

Description. Grains tetraporate, occasionally triporate or pentaporate; in polar view circular, 27–34 μm, av 31 μm in diam; in meridional view elliptical, 20–26 μm, av 23 μm wide; pores not prominent, 2–3 μm in diam, slight annular thickening, grains with 4 and 5 pores unevenly distributed around the equator; sculpturing scabrate; ektexine twice as thick as the endexine, wall 1.5 μm thick; structure tectate.

SEM: Sculpturing scabrate, with arcus joining the pores.

Flowering. April to June.

Native to. Eastern North America.

Distribution. Ontario to Nova Scotia.

Notes. In sandy soils where sweet fern is often found, it is known to shed large amounts of pollen. Their role in causing hay fever is not fully understood (Wodehouse 1971).

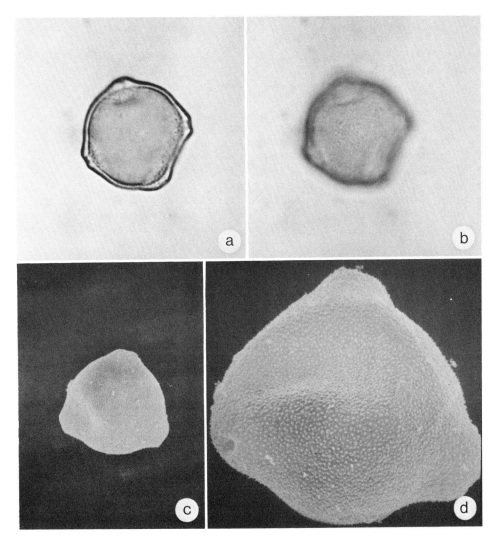

Fig. 107. *Comptonia peregrina*. *a–b*, light microscope, × 1000, *a*, polar view, equatorial section, *b*, surface; *c–d*, SEM, *c*, × 1400, *d*, × 3500.

Pollen Key to the Three Species of *Fraxinus* (Ashes)

Grains mostly tetracolpate

A. Wall 1.5 μm thick .... ***Fraxinus americana***
    F. nigra

A. Wall 1.0 μm thick .... ***F. pennsylvanica***

OLEACEAE

Fraxinus americana L.   White ash, Canadian white ash, American ash.

Description.   Grains tetracolpate, occasionally pentacolpate, rarely tricolpate (5:2:1 ratio); isopolar with colpi equally distributed around the grain; polar area index 0.48–0.59; in polar view tetracolpate grains, 23–29 μm, av 26 μm in diam; in meridional view, 21–27 μm, av 24 μm wide; pentacolpate grains larger than tetracolpate grains, in polar view, av 34 μm in diam; tricolpate grains smaller than tetracolpate grains, in polar view av 23 μm in diam; sculpturing microreticulate; endexine twice as thick as the ektexine, wall 1.5 μm thick with the thickest part at the polar ends; structure intectate, baculate.

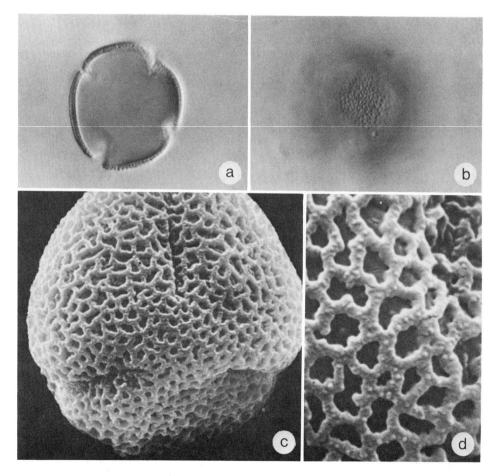

Fig. 108. *Fraxinus americana. a–b*, light microscope (interference contrast), × 1000, *a*, polar view, equatorial section, *b*, surface; *c–d*, SEM, *c*, × 4350, *d*, × 8550.

SEM: Microreticulate elements on an otherwise smooth endexine; muri rounded with microscabrate particles; luminae irregular shaped, 0.2–0.9 μm in diam.

Flowering. End of May to the middle of June.

Native to. Eastern North America.

Distribution. Ontario to Nova Scotia.

Notes. Ash pollen is caught on exposed slides in large amounts and is generally included among the pollen grains with hay fever possibilities. However, the cases of hay fever directly due to this pollen seem to be extremely rare.

OLEACEAE

Fraxinus nigra Marsh. Black ash, swamp ash, water ash, brown ash, hoop ash.

Description. Grains similar to those of *Fraxinus pennsylvanica* except for a thicker wall of 1.5 μm, endexine 0.5 μm and ektexine 1.0 μm thick; baculate particles thicker and less densely spaced.

SEM: Sculpturing microreticulate elements; muri, thickly joined baculate particles often thicker in diam than the enclosed luminae, irregularly spaced microverrucate particles on top of the muri; luminae irregular in shape, 0.3–0.9 μm in diam.

Flowering. May and June.

Native to. Eastern North America.

Distribution. Southeastern Manitoba to Newfoundland.

Notes. Cases of hay fever caused by the pollen of black ash are unknown.

(Fig. 109 overleaf)

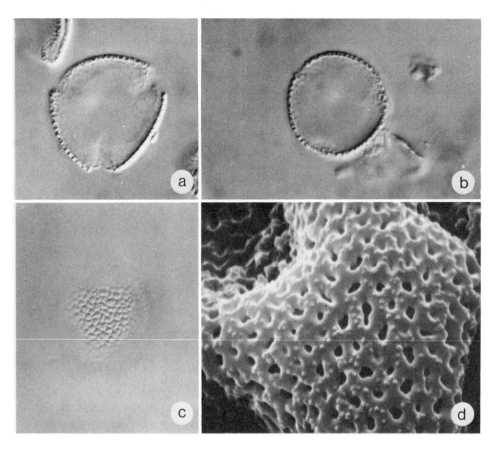

Fig. 109. *Fraxinus nigra.* a–c, light microscope (interference contrast), × 1000, a–b, polar view, equatorial section, c, surface; d, SEM, × 8750.

OLEACEAE

Fraxinus pennsylvanica Marsh.   Red ash, green ash, soft ash, river ash, brown ash, rim ash.

Description.   Grains similar to those of *Fraxinus americana* and *F. nigra* except wall only 1.0 μm thick with the endexine about 0.7 μm and the ektexine about 0.3 μm.

SEM: Sculpturing microreticulate elements on an otherwise smooth surface; muri, fused microbaculate segments densely packed, often thicker than the luminae enclosed within the muri; luminae irregular in shape, 0.15–0.70 μm in diam.

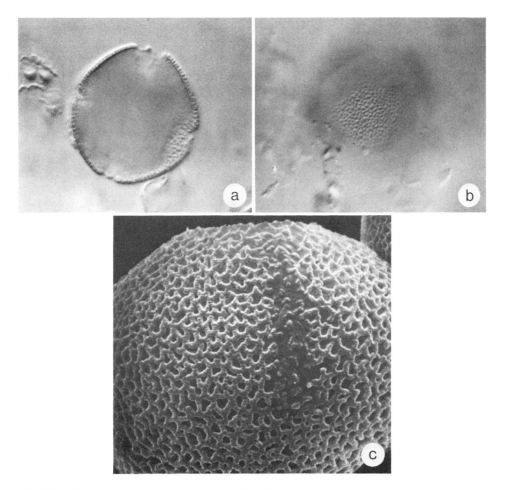

Fig. 110. *Fraxinus pennsylvanica. a–b*, light microscope (interference contrast), *a*, polar view, equatorial section, *b*, surface; *c*, SEM, × 4350.

Flowering. May and June.

Native to. North America.

Distribution. Saskatchewan to Nova Scotia.

Notes. Cases of hay fever caused by the red ash are unknown.

Pollen Key to Eight Species in the Plantaginaceae (Plantain Family)

A. Pores with a distinct operculum .... ***Plantago lanceolata***

A. Pores with no distinct operculum

    B. Pores 4–9

        C. Grains less than 23 μm in diam

            D. Wall thickness about 1 μm, pores about 3 μm in diam .... ***P. major***

            D. Wall thickness about 2 μm, pores more than 4 μm in diam .. ***P. canescens***

        C. Grains more than 23 μm in diam

            E. Pores with jagged margins, no annulus

                F. Verrucae about 2.5 μm in diam under SEM .... ***P. rugelii***

                F. Verrucae about 1 μm in diam under SEM .... ***P. macrocarpa***

            E. Pores with entire or slightly irregular margins, with or without annulus

                G. Verrucae about 1 μm in diam, pores slightly annulate under SEM .... ***P. maritima***

                G. Verrucae about 1.5 μm in diam, no annulus under SEM .... ***P. eriopoda***

    B. Pores 12–14 .... ***Littorella americana***

PLANTAGINACEAE

Littorella americana Fernald   American littorella, littorella.

Description. Grains periporate, 12–14 pores, granules within pores, jagged margins, 2.5–3.0 μm in diam, no annulus; grains spheroidal, occasionally ovoidal, 31.5–36.5 μm, av 34.0 μm in diam; sculpturing verrucate, surface covered with many microechinate granular particles; ektexine and endexine over 3 μm thick; structure tectate.

    SEM: Verrucae about 2 μm in diam.

Flowering. Latter part of June to August.

Native to. Eastern North America.

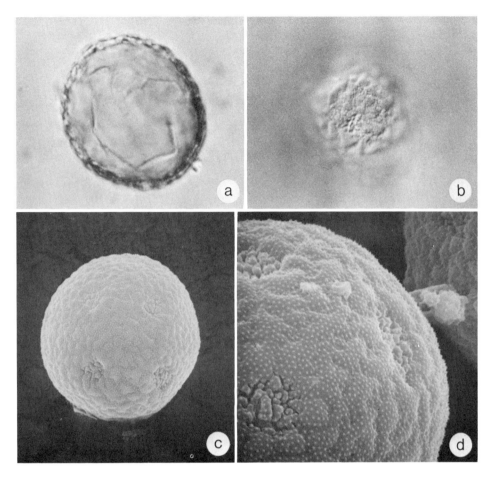

Fig. 111. *Littorella americana.* a-b, light microscope (interference contrast), × 1000, a, equatorial section, b, surface; c-d, SEM, c, × 1500, d, × 3700.

Distribution. Ontario to Newfoundland, common along shorelines of rivers and lakes in Nova Scotia and other parts of the Maritime Provinces.

Notes. Although there appears to be considerable pollen shed from the flowers, there is no available information that the grains cause hay fever.

PLANTAGINACEAE

Plantago canescens Adams   Siberian plantain, Arctic plantain.

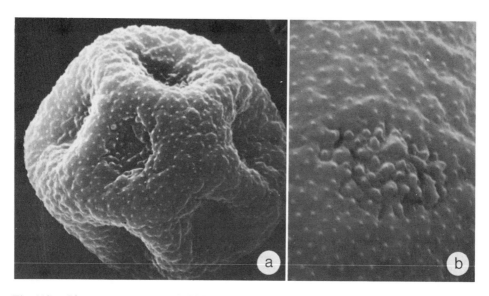

Fig. 112.   *Plantago canescens. a-b*, SEM, *a*, × 4000, *b*, × 7700.

Description.   Grains periporate, 4-6 (-7) pores, granules within the pores, jagged margins, 4-5 μm in diam, no annulus; grains spheroidal, rarely irregular, 18.5-21.0 μm, av 20.0 μm in diam; sculpturing verrucate, surface covered with many microechinate granular particles; ektexine and endexine about 2 μm thick; structure tectate.

   SEM: Grains similar to the pollen of *P. rugelii*; verrucae about 2 μm in diam.

Flowering.   Mid-May to July.

Native to.   North America.

Distribution.   Southwestern Alberta to the Yukon and the Northwest Territories.

Notes.   Although this species sheds considerable pollen (Bassett and Crompton 1968), it is not known to be important in causing hay fever.

PLANTAGINACEAE

Plantago eriopoda Torr.   Saline plantain, redwood plantain.

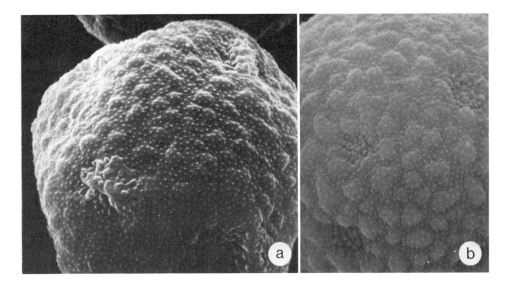

Fig. 113.   *Plantago eriopoda. a–b*, SEM, *a*, × 4000, *b*, × 4500.

Description.   Grains periporate, (5–) 6–8 pores, granules within pores, margins of pores nearly entire, about 3 µm in diam, no annulus; grains spheroidal, rarely irregular, 23.0–25.5 µm, av 24.0 µm in diam; sculpturing verrucate, surface covered with many microechinate granule particles; ektexine and endexine about 1.5 µm thick; structure tectate.

SEM: Microechinate spines covering the whole surface; pores depressed, surface covered with wartlike particles and microechinate spines as in *P. rugelii*; verrucae about 1.5 µm in diam.

Flowering.   Early May to July.

Native to.   North America.

Distribution.   Yukon, Northwest Territories, British Columbia, Prairie Provinces, and along the shores of the St. Lawrence River from Quebec City to the Gaspé Peninsula and Anticosti Island (Bassett 1973).

Notes.   It is doubtful if enough pollen is shed by this species to cause hay fever.

PLANTAGINACEAE

Plantago lanceolata L. English plantain, buckhorn, narrow-leaved plantain, ripplegrass, ribgrass, ribwort, black plantain.

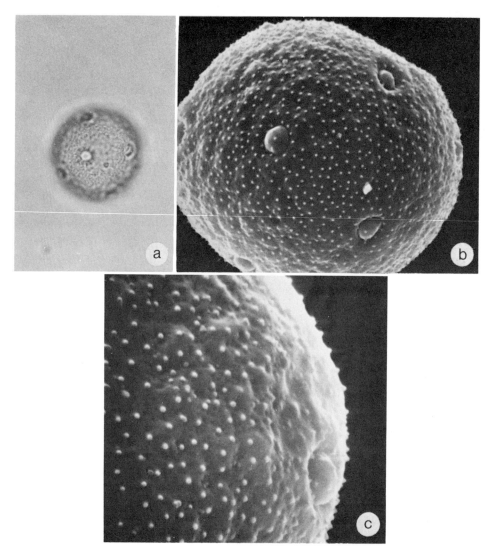

Fig. 114. *Plantago lanceolata. a*, surface, pores, annulus, and operculum, light microscope, × 1000; *b–c*, SEM, *b*, × 4000, *c*, × 8050.

Description. Grains periporate, 8-14 pores, operculate, margins of pores entire, pores 2.0-2.5 μm in diam, annulus distinct and about 1 μm wide; grains spheroidal, rarely irregular, 23.0-27.5 μm, av 25.5 μm in diam; sculpturing slightly verrucate, surface covered with many microechinate granular particles; ektexine and endexine about 1 μm thick; structure tectate.

SEM: Microechinate spines covering the operculum and the whole surface of the grain; annulus more pronounced than in *P. maritima*; verrucae about 1 μm in diam.

Flowering. Early June to October, with peak the end of June and early July.

Native to. Europe and Asia.

Distribution. British Columbia and from Ontario to Newfoundland, common along the southern coastline of British Columbia, southern Ontario, Quebec, and along the Atlantic coastline.

Notes. The pollen is undoubtedly an important cause of hay fever especially in areas where the species is common.

PLANTAGINACEAE

Plantago macrocarpa Cham. & Schlect.   Alaska plantain, North Pacific plantain.

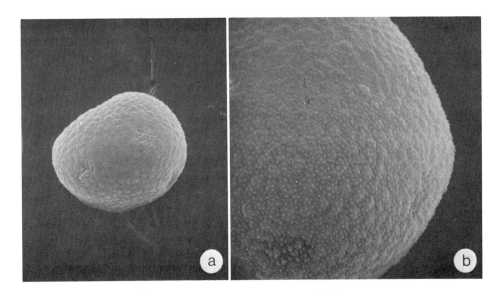

Fig. 115. *Plantago macrocarpa.* a-b, SEM, a, × 1550, b, × 3800.

Description. Grains periporate, 4–6 pores, granules within pores, pores with jagged margins, about 3 μm in diam, no annulus; grains spheroidal, rarely irregular, 26.0–28.5 μm, av 27.0 μm in diam; sculpturing slightly verrucate, surface covered with many microechinate granular particles; ektexine and endexine about 1 μm thick; structure tectate.

SEM: Grains similar to the pollen of *P. rugelii*; verrucae about 1 μm in diam.

Flowering. The latter part of May to early July.

Native to. North America.

Distribution. The western coastline of British Columbia to Alaska.

Notes. Although pollen has been collected on exposed slides (Bassett and Crompton 1966), it has not been in sufficient quantities to cause hay fever.

PLANTAGINACEAE

Plantago major L. Common plantain, broad-leaved plantain, dooryard plantain, whiteman's foot.

Description. Grains periporate, (3–) 4–6 (–7) pores with jagged margins, granules within pores, about 3 μm in diam, no annulus, grains spheroidal, rarely irregular, 21.0–24.5 μm, av 22.5 μm in diam; sculpturing verrucate, surface covered with many microechinate granular particles; ektexine and endexine about 1 μm thick; structure tectate.

SEM: Microechinate spines covering the whole surface; pores depressed, surface covered with wartlike particles and microechinate spines; verrucae about 1.5 μm in diam.

Flowering. Mid-June to October.

Native to. Europe and Asia.

Distribution. Northward to the tree line, generally in areas disturbed by man such as cultivated fields, lawns, roadsides, and waste places in all provinces, not common in shaded areas or in areas continually wet during the growing season (Bassett 1973).

Notes. Common plantain sheds so little pollen that no matter how abundant the plants may be, it can never be considered a major factor in causing hay fever. This species contains the same antigenic factor as English plantain (Wodehouse 1971).

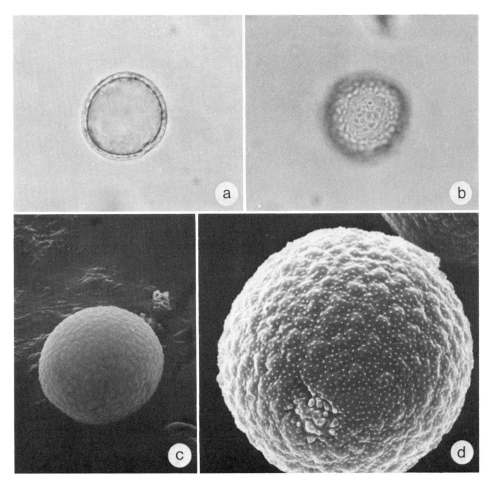

Fig. 116. *Plantago major.* a–b, light microscope, × 1000, a, equatorial section, b, surface; c–d, SEM, c, × 1800, d, × 4000.

PLANTAGINACEAE

Plantago maritima L.   Seaside plantain, seashore plantain, rushlike plantain.

Description. Grains periporate, 4–6 (–8) pores, granules within pores, margins of pores slightly irregular or nearly entire, about 2.5–3.0 μm in diam, with or without an annulus, annulus about 1 μm wide; grains spheroidal, rarely irregular, 26–28 μm, av 27 μm in diam; sculpturing slightly verrucate, surface covered with many microechinate granular particles; ektexine and endexine about 1 μm thick; structure tectate.

223

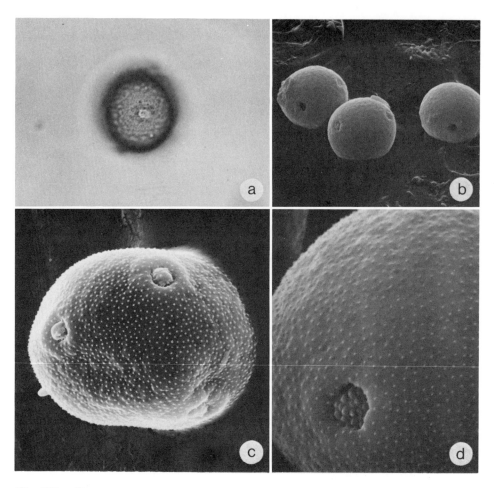

Fig. 117. *Plantago maritima*. *a*, surface and pore, light microscope, × 1000; *b–d*, SEM, *b*, × 800, *c*, × 2000, *d*, × 3800.

SEM: Microechinate spines covering the whole surface; pores slightly annulate, pore surface with larger echinate particles than on overall grain surface; verrucae about 1 μm in diam.

Flowering. The latter part of May through to July.

Native to. North America.

Distribution. Along the Pacific and Atlantic coastlines and the shorelines of Hudson, James, and Ungava bays.

Notes. Insufficient pollen from seaside plantain has been caught on exposed slides to suggest that it might be important in causing hay fever.

PLANTAGINACEAE

Plantago rugelii Decne. Rugel's plantain, purple-stemmed plantain, pale plantain, broad-leaved plantain, silk plantain, whiteman's foot.

Description. Grains periporate, (5-) 6-8 (-9) pores, granules within the pores, pores with jagged margins, about 3 μm in diam, no annulus; grains spheroidal, rarely irregular, 23-27 μm, av 25 μm in diam; sculpturing verrucate, surface covered with many microechinate granular particles; ektexine and endexine about 2 μm thick; structure tectate.

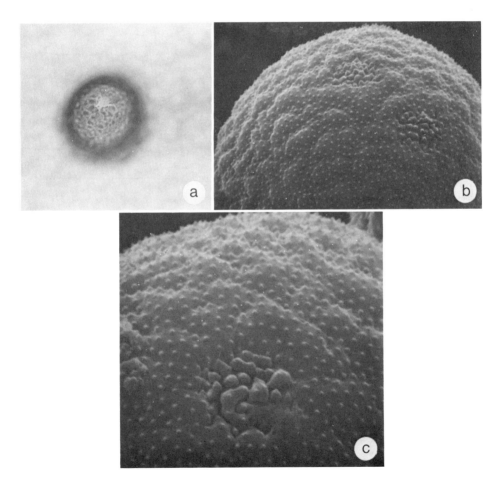

Fig. 118. *Plantago rugelii. a,* surface and pore, light microscope, × 1000; *b-c,* SEM, *b,* × 3850, *c,* × 7850.

225

SEM: Microechinate spines covering the whole surface as in *P. major*; pores depressed, surface composed of verrucate particles and microechinate spines; verrucae about 2.5 μm in diam.

Flowering. Early part of June to October.

Native to. North America.

Distribution. Ontario to Nova Scotia, most common in southern Ontario.

Notes. A few pollen grains from Rugel's plantain have been caught on exposed slides. It can be considered a minor factor in causing hay fever.

PLATANACEAE

Platanus occidentalis L.   Sycamore, plane tree, buttonball, buttonwood.

Description. Grains tricolpate, occasionally tetracolpate, oblate, 17–20 × 15–18 μm, av 18.5 × 16 μm; colpi margins distinct except at the polar ends, verrucate particles; sculpturing reticulate; ektexine and endexine of equal thickness, wall 1 μm thick; structure baculate tectate.

SEM: Sculpturing reticulate with irregular shaped luminae; verrucate-rugulate particles covering the colpi.

Flowering. Latter part of May and early June.

Native to. Eastern North America.

Distribution. Southern Ontario.

Notes. The sycamore and the London plane, *Platanus acerifolia* (Ait.) Willd. are often cultivated in urban areas. Wodehouse (1971) stated that the trees shed large amounts of pollen, which can be as important as oak pollen in causing hay fever.

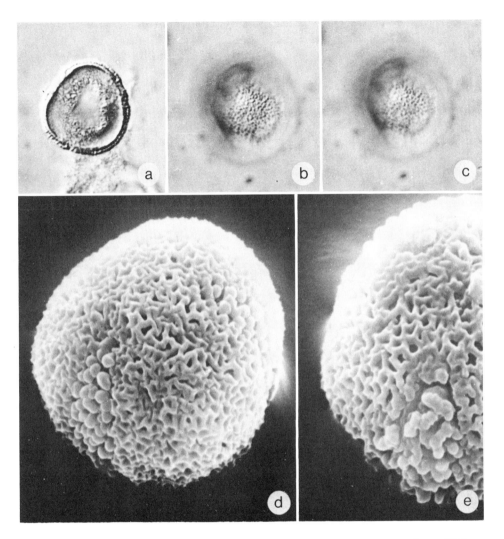

Fig. 119. *Platanus occidentalis. a–c*, light microscope (interference contrast), × 1000, *a*, polar view, equatorial section; *b–c*, surface; *d–e*, SEM, × 5000.

POLYGONACEAE

Eriogonum flavum Nutt. Yellow umbrella plant, umbrella plant.

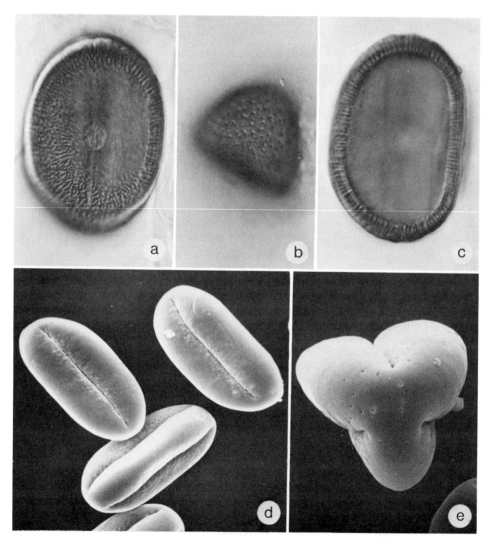

Fig. 120. *Eriogonum flavum.* a–c, light microscope (interference contrast), × 1000, a, view of colpi and pore, b, surface of polar end, c, equatorial section; d–e, SEM, d, × 730, e, × 1500.

Description. Grains tricolporate; polar area index appears large (difficult to calculate as the grains lie on their equatorial surface); in polar view circular; in meridional view prolate, 30–36 μm, av 33 μm wide × 42–54 μm, av 48 μm long; pores not prominent, generally circular or slightly oblong, 4–6 μm in diam, formed by a break in the endexine and not shown on the exterior of the grain; sculpturing, tuberculate processes evenly distributed over a layer of smooth ektexine surface based on long micropilae; endexine three times as thick as the ektexine, which is a solid surface with slightly raised baculate pilae, wall 3–5 μm thick, very thick at the polar ends, endexine of tightly fused striate pilae based on a slight tectum; structure tectate.

SEM: Sculpturing, smooth exine except for shallowly reticulate furrow margins, polar ends irregular microperforated.

Flowering. July.

Native to. Western North America.

Distribution. Southern British Columbia, Alberta, Saskatchewan, and southwestern Manitoba.

Notes. Five *Eriogonum* spp. occur in Western Canada. None of these taxa sheds enough pollen to be important in causing hay fever.

POLYGONACEAE

Oxyria digyna (L.) Hill. Mountain sorrel, sorrel.

Description. Grains tricolporate, occasionally tetracolporate; polar area index 0.32; in polar and equatorial views circular to oblate, 21–25 μm, av 23 μm wide × 21–25 μm, av 24 μm long; pores indistinct, circular to oblong, 3 μm in diam; sculpturing, uneven granular processes; ektexine and endexine thickness indistinct, wall 1 μm thick; structure tectate.

SEM: Grains similar to those of *Rumex obtusifolius.*

Flowering. Late May to August depending on elevation and latitude.

Native to. Western North America.

Distribution. Yukon, Northwest Territories, British Columbia, Alberta, Quebec, Nova Scotia, and Labrador.

Notes. When shed in large amounts, the pollen from this species may cause hay fever.

(Fig. 121 overleaf)

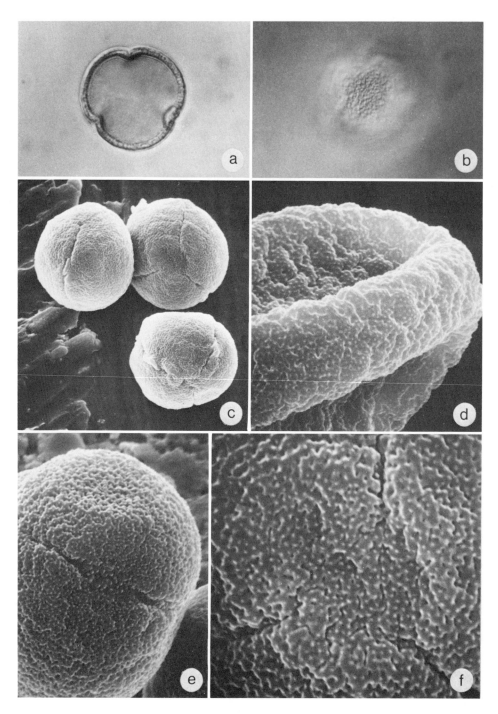

Fig. 121. *Oxyria digyna.* a–b, light microscope (interference contrast), × 1000, a, polar view, equatorial section, b, surface of polar end; c–f, SEM, c, × 1500, d, × 4000, e, × 3650, f, × 7350.

POLYGONACEAE

Polygonum lapathifolium L.   Pale smartweed, pale persicaria, dock-leaved smartweed.

Description.   Grains periporate, spheroidal, 30–50 μm, av 40 μm in diam; pores 13–20, circular, 3 μm in diam, without microgranular coverings; sculpturing reticulate, muri acute, supported by a double flanking row of micropilae, some luminae small and round, 3 μm in diam without enclosed ornamentation, others irregular in shape, 4–8 μm in diam with microgemmate or microverrucate particles that could be pore coverings; ektexine more than twice as thick as the endexine, wall 4 μm thick; structure tectate, with double row of baculate rods fused to form echinate structure in optical section, small verrucate particles between these structures might be pore coverings.

SEM: Grains same as under light microscope; loosely attached, complex, reticulate exine, with acute topped muri supported by a double row of pilae; when reticulum removed mechanically many pores visible covered with a varying number of microverrucate or microgemmate particles.

Flowering.   July, August, and September.

Native to.   Europe and Asia.

Distribution.   British Columbia to Newfoundland.

Notes.   Very little pollen from this species is shed into the air. The pollen morphology in this genus is very variable (Hedberg 1946) and requires further study.

(Fig. 122 overleaf)

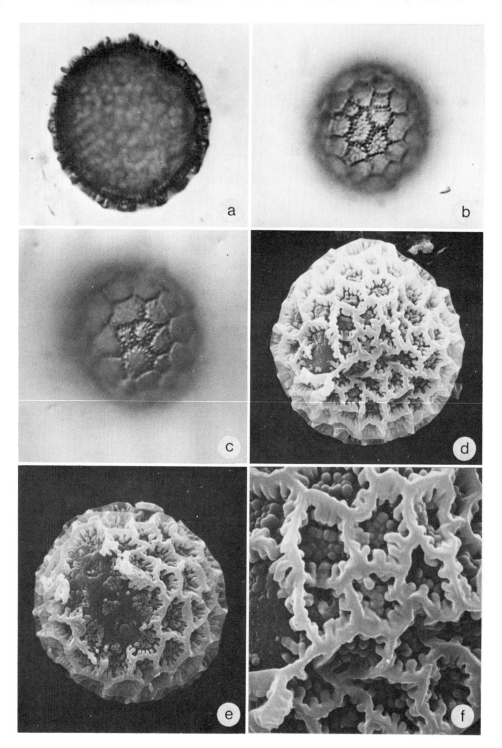

Fig. 122. *Polygonum lapathifolium. a–c*, light microscope (interference contrast), × 1000, *a*, equatorial section, *b–c*, surface; *d–g*, SEM, *d–e*, × 1500, *f*, × 3700.

Pollen Key to Four Species of *Rumex* (Docks)

A. Grains mostly pericolporate, averaging 33 μm in diam .... ***Rumex orbiculatus***

A. Grains stephanocolporate, tetracolporate, tricolporate, rarely pericolporate, averaging less than 30 μm in diam

    B. Grains averaging less than 23 μm in diam .... ***R. acetosa***

    B. Grains averaging more than 23 μm in diam .... ***R. acetosella***
                                                                                                                     R. obtusifolius

POLYGONACEAE

Rumex acetosa L.   Garden sorrel, garden dock.

Description. Grains tricolporate, tetracolporate, rarely pericolporate; polar area index 0.6-0.7; grains spheroidal, 19-23 μm, av 21 μm in diam; colpi slightly intruding of various lengths and positions on the grain surface, none completely encircling the grain; all other pollen characteristics similar to those of *R. acetosella.*

SEM: Sculpturing, small circular to irregular shaped perforations, surrounded by thick muri upon which distinct, densely spaced, microscabrate processes are based.

Flowering.   June, July, and August.

Native to.   Europe.

Distribution.   All provinces, fairly common in southeastern Quebec.

Notes. Where garden sorrel is common the pollen is shed in large amounts, but is apparently of minor importance in causing hay fever.

(Fig. 123 overleaf)

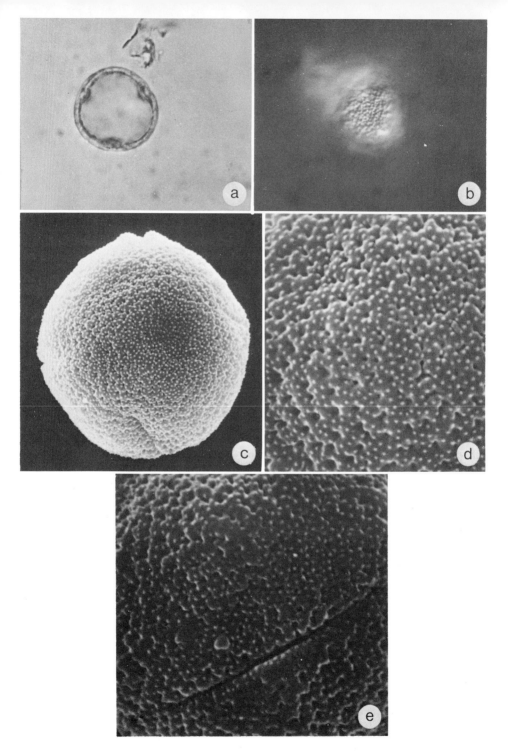

Fig. 123. *Rumex acetosa*. a–b, light microscope, × 1000, a, polar view, equatorial section, b, surface, under interference contrast; c–e, SEM, c, × 3800, d–e, × 7300.

POLYGONACEAE

Rumex acetosella L. Sheep sorrel, field sorrel, horse sorrel, red sorrel, sourgrass.

Description. Grains stephanocolporate, tetracolporate, rarely tricolporate or pericolporate, spheroidal, 22–30 μm, av 25.5 μm in diam; colpi slightly intruding of various lengths and positions on the surface, some colpi meet with other colpi encircling the grain; pores indistinct when acetolyzed and mounted in silicone oil, distinct when mounted in glycerine jelly due to expansion, pores oriented within and in the direction of the colpi, oblong 1–3 μm wide, 3–6 μm long, bracketed with a slight margo along the colpi edges; sculpturing reticulate with comparatively thick muri containing small luminae; wall 1.5 μm thick; structure tectate with simple, densely spaced columellae.

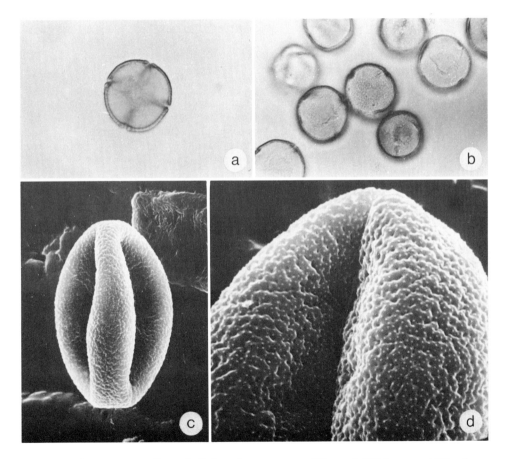

Fig. 124. *Rumex acetosella.* a–b, light microscope, × 500; c–d, SEM, c, × 1500, d, × 3650.

SEM: Sculpturing similar to the grains of *R. acetosa* except microscabrate processes less densely spaced.

Flowering. June, July, and August.

Native to. Europe and Asia.

Distribution. All provinces, very common in southern British Columbia and Eastern Canada.

Notes. Sheep sorrel is known to shed large quantities of pollen, but the allergenic activity of the pollen is evidently very low (Feinberg 1946).

POLYGONACEAE

Rumex obtusifolius L. Broad-leaved dock, bitter dock, blunt-leaved or red-veined dock.

Description. Grains stephanocolporate, rarely tetracolporate, occasionally pericolporate, spheroidal, 24–34 µm, av 29 µm in diam; colpi slightly intruding, various lengths and positions on the surface bracketed with a prominent margo; pores indistinct when acetolyzed and mounted in silicone oil, distinct when mounted in glycerine jelly, oriented in the direction of the colpi, oblong, 1–3 µm wide, 6–10 µm long; sculpturing microreticulate with comparatively thick muri containing small luminae; ektexine and endexine of equal thickness except at the pore area where the endexine thickens to form an interior margo; structure tectate with simple densely spaced columellae.

SEM: Sculpturing mostly microrugulate, sometimes luminae small circular perforations, muri rounded and form the base for evenly distributed microscabrate sculpturing elements.

Flowering. July and August.

Native to. Europe.

Distribution. British Columbia and from Ontario to Newfoundland.

Notes. The pollen is of minor importance in causing hay fever.

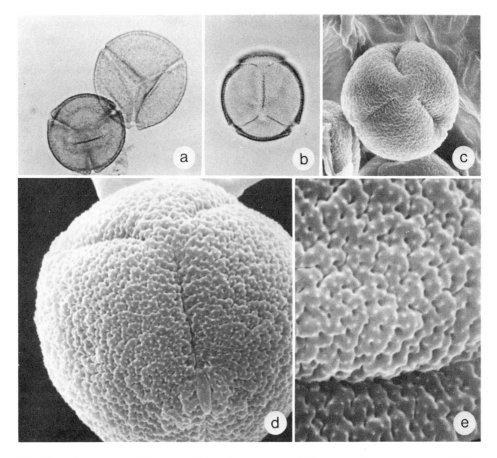

Fig. 125. *Rumex obtusifolius. a–b*, light microscope, × 1000, equatorial sections; *c–e*, SEM, *c*, × 1650, *d*, × 4200, *e*, × 8750.

POLYGONACEAE

Rumex orbiculatus Gray   Water dock, swamp dock.

Description. Grains pericolporate, occasionally syncolporate, rarely tricolporate or tetracolporate, spheroidal, 28–37 μm, av 33 μm in diam; colpi slightly intruding of various lengths and positions on the surface, inconspicuous margo; pores oblong, 2–3 μm wide, 7–11 μm long; sculpturing microreticulate; ektexine thinner than the endexine, wall 1 μm thick; structure tectate.

SEM: Sculpturing, circular microfoveolate pits form a very shallow pseudoreticulum over the surface, muri of this pseudoreticulum form the base for evenly distributed microscabrate sculpturing elements.

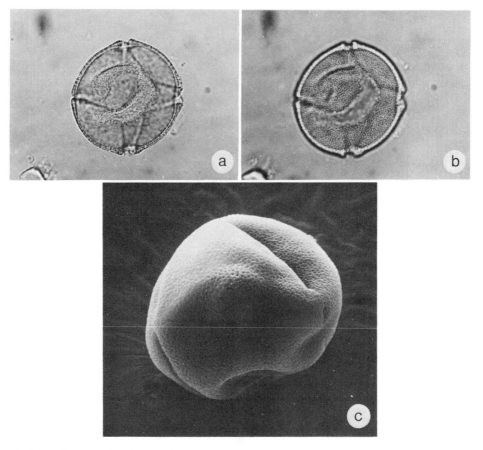

Fig. 126. *Rumex orbiculatus. a–b*, light microscope, × 1000, *a*, polar view, equatorial section, *b*, surface; *c*, SEM, × 1600.

Flowering. Latter part of July and August.

Native to. North America.

Distribution. Mackenzie Mountains, Yukon, and from Alberta to Newfoundland; more common in southeastern Canada.

Notes. The pollen seems to be of minor importance in causing hay fever.

RANUNCULACEAE

Thalictrum dasycarpum Fisch. & Lall.   Meadow rue.

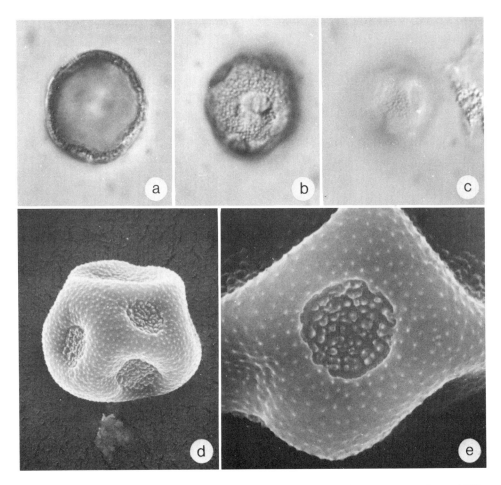

Fig. 127. *Thalictrum dasycarpum*. a–c, light microscope (interference contrast), × 1000, a, equatorial section, b–c, different focus of the pores and surface; d–e, SEM, d, × 1700, e, × 4300.

Description. Grains periporate, circular, 23–27 μm, av 25 μm in diam; pores 7–20 μm, av 14 μm, circular, 7–10 μm in diam, somewhat intruding, no annulus, pore membrane flecked with many small microgranules; sculpturing microechinate; ektexine and endexine of equal thickness, wall 2 μm thick; structure tectate.

SEM: Sculpturing microechinate with spines scattered evenly over an otherwise smooth surface; pores somewhat depressed with various sizes of microgranules or verrucate particles topped by microechinate spines, spaced out showing the endexine beneath.

Flowering. Late May and June.

Native to. North America.

Distribution. Alberta to Ontario.

Notes. Occasionally a few pollen grains of *Thalictrum* spp. are caught on exposed slides, but they are not known to cause hay fever.

ROSACEAE

Sanguisorba canadensis L.   Canadian burnet, burnet.

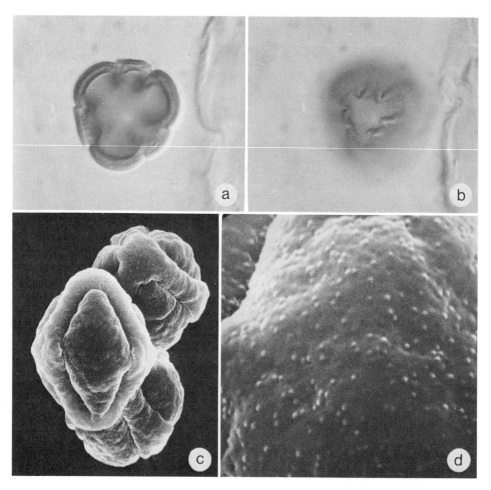

Fig. 128. *Sanguisorba canadensis. a–b,* light microscope (interference contrast), × 1000, *a,* polar view, equatorial section, *b,* polar surface; *c–d,* SEM, *c,* × 1550, *d,* × 7350.

Description. Grains stephanocolporate, prolate, 24–30 μm wide × 31–40 μm long, av 27 × 34 μm; polar area index 0.6; in meridional view each pore bordered by two colpi, between colpi a solid diamond-shaped plate of exine; pores oriented lengthwise along the equator with the width oriented towards the poles; in polar view pores 3 μm wide; in meridional view pores 6 μm long; pores with an interior, endexinous, thickened margin.

SEM: Pores not clearly visible on material examined, appear as slight protuberances on the grain equator; exine surface undulating, microgranulate processes evident at high magnification (6000–8000 ×).

Flowering. August to mid-September.

Native to. North America.

Distribution. British Columbia, Quebec, New Brunswick, Nova Scotia, Newfoundland, and Labrador.

Notes. A few pollen grains from this species have been caught on exposed slides. Pollen grains from *Pyrus, Malus,* and other genera in the Rosaceae that have been caught on exposed slides are very rare. Pollen grains of *Pyrus, Malus,* and *Prunus* spp. are similar to those in the maple family, Aceraceae. Adams and Morton (1974) have excellent photographs taken with the SEM of several genera in the rose family.

SALICACEAE

Populus grandidentata Michx. Largetooth aspen, largetooth poplar, bigtooth aspen.

Description. Grains inaperturate, spheroidal, 25.5–30.0 μm, av 27.5 μm in diam; sculpturing scabrate; ektexine and endexine about 0.5 μm thick; structure intectate.

SEM: Surface covered densely with regularly or slightly irregularly small, mostly rounded processes.

Flowering. Around April 24 with the peak period shortly afterwards at Ottawa (Bassett et al. 1961), in the extreme southern section of Ontario, 1 or 2 weeks earlier.

Native to. Eastern North America.

Distribution. Eastern Manitoba to Nova Scotia.

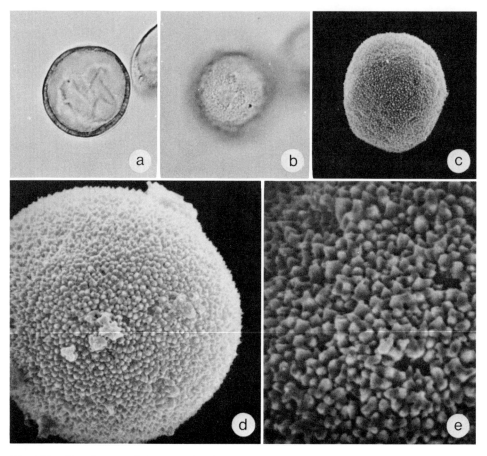

Fig. 129. *Populus grandidentata a–b*, light microscope (interference contrast), × 1000, *a*, equatorial section, *b*, surface; *c–e*, SEM, *c*, × 1600, *d*, × 3850, *e*, × 7350.

Notes. The pollen from the largetooth aspen has to be shed in large amounts to cause hay fever. Pollen from *P. balsamifera* L., *P. deltoides* Bartr., and *P. tremuloides* Michx. are similar in ornamentation and size, but that of *P. balsamifera* tends to be smaller (Adams and Morton 1972). In cities, towns, and villages many cultivated species of *Populus* have been planted including the silver poplar, *P. alba* L., and lombardy poplar, *P. nigra* L. var. *italica* DuRoi. They can shed large amounts of pollen that can cause hay fever.

SALICACEAE

Salix discolor Muhl.   Pussy willow, willow.

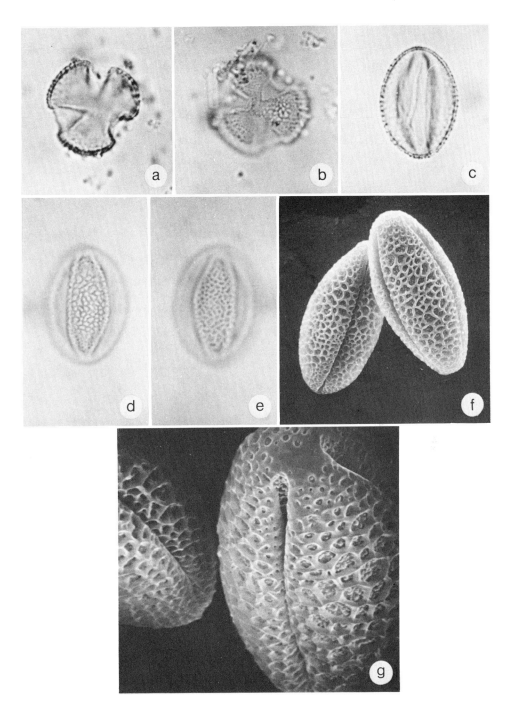

Fig. 130. *Salix discolor. a–e*, light microscope, × 1000, *a*, polar view, equatorial section, *b*, polar surface, *c*, equatorial section, *d–e*, equatorial surface at different focus; *f–g*, SEM, *f*, × 1650, *g*, × 5500.

Description. Grains tricolporate, prolate, 16–22 × 27–30 μm, av 19.0 × 28.5 μm; polar area index 0.23; colpi open, intruding; sculpturing reticulate, microreticulate close to the edges of the colpi; ektexine twice as thick as the endexine, wall 1 μm thick; structure tectate, baculate.

SEM: Sculpturing reticulate with verrucate particles within the luminae; muri acute at top; luminae irregular, 0.3–1.0 μm in diam.

Flowering. April and early May.

Native to. North America.

Distribution. British Columbia to Newfoundland and in the Yukon and Northwest Territories.

Notes. The willows are primarily insect-pollinated, but a number of pollen grains have been caught on exposed slides at stations throughout Canada. Durham (1951) stated that the pollen grains of the willows are only of minor importance in causing hay fever.

TILIACEAE

Tilia americana L.   Basswood, linden, lime, whitewood.

Description. Grains tricolporate; polar area index 0.65–0.75; in polar view mostly circular, 37–45 μm, av 41 μm in diam; in meridional view isopolar, 23–31 μm, av 27 μm wide; colpi elliptical in shape with irregular margins, 4–6 μm wide × 10–16 μm long; sculpturing irregular microreticulate; ektexine and endexine of equal thickness except near the colpi where endexine 4–7 μm thick, otherwise wall 3 μm thick; structure tectate, columellae of branched micropilae.

SEM: Sculpturing irregular microreticulate with luminae up to 1 μm in diam; irregular colpi margins.

Flowering. Late June and July.

Native to. Eastern North America.

Distribution. Eastern Manitoba to western New Brunswick.

Notes. Wodehouse (1935) mentioned that although basswood is insect-pollinated it sheds large amounts of pollen. It is not an important factor in hay fever. The pollen is often found in postglacial silts.

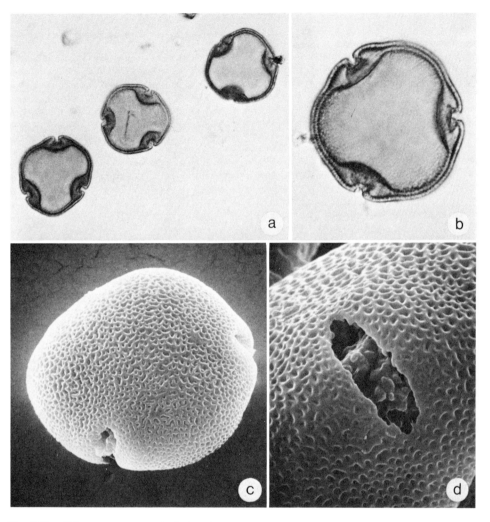

Fig. 131. *Tilia americana.* a–b, light microscope, a, polar view, equatorial section, × 400, b, × 1000; c–d, SEM, c, × 1800, d, × 4500.

TYPHACEAE

Typha angustifolia L.   Narrow-leaved cattail, cattail.

Description.   Grains monoporate, rarely united in diads, triads, or tetrads; grains subspheroidal, mostly free, 19–27 µm, av 24 µm in diam; pores 3.0–4.5 µm in diam, circular in outline, margins irregular and not clearly defined because they merge with the sculpturing pattern; sculpturing reticulate supported by baculate micropilae; ektexine twice as thick as the endexine, wall 2 µm thick; structure tectate.

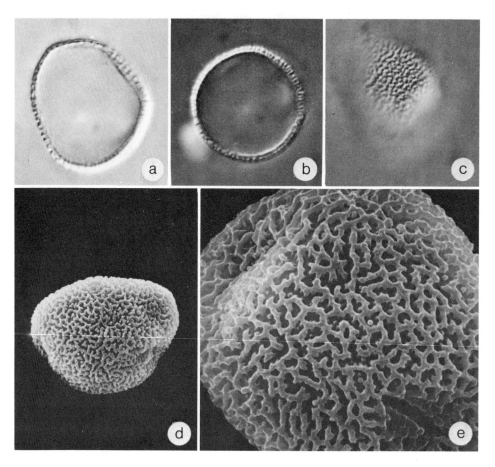

Fig. 132. *Typha angustifolia.* a–c, light microscope (interference contrast), × 1000, a–b, equatorial section, c, surface; d–e, SEM, d, × 1500, e, × 3600.

SEM: Sculpturing reticulate supported by single or double micropilae; verrucate particles within some luminae, luminae irregular, 1–5 μm long.

Flowering. Latter part of May, June, and early July.

Native to. North America.

Distribution. Manitoba to Nova Scotia.

Notes. Although large amounts of pollen are shed by this species, only occasionally has it been known to cause hay fever (Wodehouse 1971).

TYPHACEAE

Typha latifolia L.   Broad-leaved cattail, common cattail.

Description.   Grains mostly united in tetragonal tetrads in one plane; single grains rare, monoporate, overall size 36–45 µm, av 40 µm in diam, individual grains 18–26 µm in diam; each grain with a single pore, normally on the distal part, pore circular in outline, 3.0–4.5 µm in diam, margins irregular and not clearly defined because they merge with the sculpturing pattern; sculpturing reticulate supported by baculate micropilae; ektexine twice as thick as the endexine, wall 1 µm thick; structure tectate.

SEM: Sculpturing reticulate, distal face degenerating into rugulate sculpturing near the proximal area; connective bands joining the grains into tetrads; luminae irregular, 0.4–0.7 µm long.

Flowering.   Latter part of May, June, and early July.

Native to.   North America.

Distribution.   All provinces and the Yukon and Northwest Territories.

Notes.   *Typha latifolia* hybridizes freely with *T. angustifolia* where their ranges overlap. The pollen of the hybrid *T. glauca* (*T. latifolia* × *T. angustifolia*) (Fig. 133) is similar to the pollen of *T. angustifolia* except that the grains are occasionally attached in diads, triads, or tetrads with many aborted and distorted grains (65%). The sculpturing is intermediate between the two parents and there are no granules within the luminae, which are 0.4–4.0 µm long. Dugle's (1972) pollen descriptions of the cattails in Manitoba are similar to those that we have found across Canada.

Only occasionally the pollen of cattails has been known to cause hay fever.

(Fig. 133 overleaf)

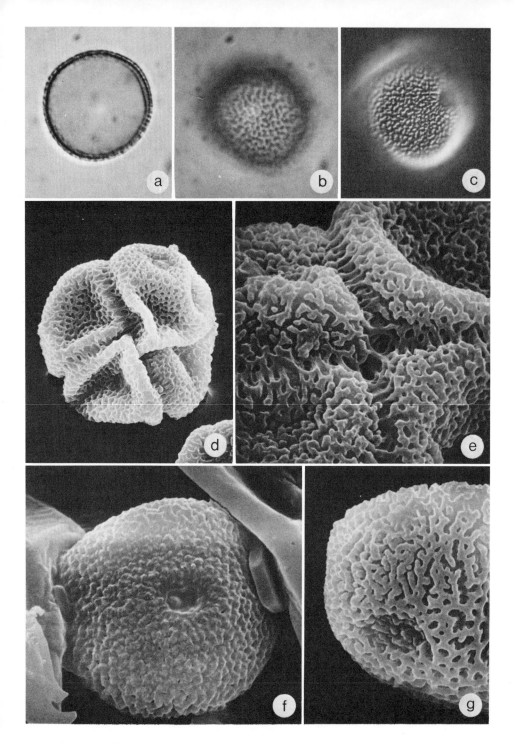

Fig. 133. *Typha latifolia.* a–c, light microscope, × 1000, a, equatorial section, b, surface, c, surface under interference contrast; d–e, SEM, d, × 1500, e, × 3700. f–g, *T. glauca*, SEM, × 3700.

ULMACEAE

Ulmus americana L. White elm, American elm, water elm, swamp elm, rock elm.

Description. Grains stephanoporate, (3–) 5–6 (–7) pores around the equator, with five pores the interval between the unpaired pore and its neighbor on one side generally visibly greater than between the others (Wodehouse 1935); if in polar view outline of grains suggests a pentagon with one side longer than the other, 30.0–33.5 μm, av 31.0 μm; pores about 4 μm in diam, aspidate; sculpturing rugulate to reticulate; endexine and ektexine about 1 μm thick; structure tectate.

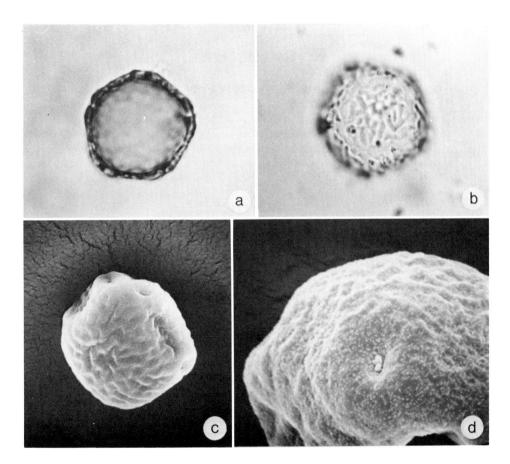

Fig. 134. *Ulmus* spp. *a–b*, light microscope, × 1000, *a*, polar view, equatorial section, *b*, surface; *c–d*, SEM, *c*, × 1600, *d*, × 3600.

SEM: Sculpturing mostly rugulate-verrucate with very fine microgranules on the surface.

Flowering. At Ottawa April 11 with the peak period of 3 or 4 days beginning 3 or 4 days later, in extreme southern part of Ontario flowering begins 1 or 2 weeks earlier than at Ottawa, in the Maritime Provinces 1 or 2 weeks later than at Ottawa.

Native to. North America.

Distribution. Eastern Saskatchewan to western Newfoundland, common primarily in southern Ontario and Quebec.

Notes. When the American elm is at maximum flowering its pollen can cause severe hay fever. Its period of flowering is only a few days in one locality.

ULMACEAE

Ulmus rubra Muhl. Slippery elm, red elm, slippery-barked elm, budded elm.

Description. Grains stephanoporate, mostly 5 pores around the equator with same arrangement as in *U. americana*; in polar view oblate, 28.9–29.0 µm, av 28.5 µm in diam; in meridional view 21–24 µm × 25.5–29.5 µm; pores elliptical, 2.0–5.5 µm in diam, aspidate; sculpturing coarser reticulate than in *U. americana*; endexine and ektexine 1–2 µm thick; structure tectate.

SEM: Sculpturing similar to *U. americana.*

Flowering. About the same time as the American elm.

Native to. Eastern North America.

Distribution. Southern Ontario to southeastern Quebec.

Notes. Pollen from this species can cause hay fever when shed in large amounts.

The pollen of the rock elm, *U. thomasii* Sarg., is very similar to that of the white elm under the light and scanning electron microscopes. Rock elm occurs from southern Ontario to the southeastern part of Quebec. When shed in large amounts, its pollen may cause hay fever.

Fig. 135. *Celtis occidentalis. a–e,* light microscope, × 1000. *a,* equatorial section, *b,* surface and pore, *c,* annulus and pore, *d,* surface, pore, and annulus, under interference contrast; *e,* surface, pore, and annulus, under phase contrast; *f,* SEM, × 3850.

ULMACEAE

Celtis occidentalis L.   Hackberry, nettle tree, sugarberry, bastard elm.

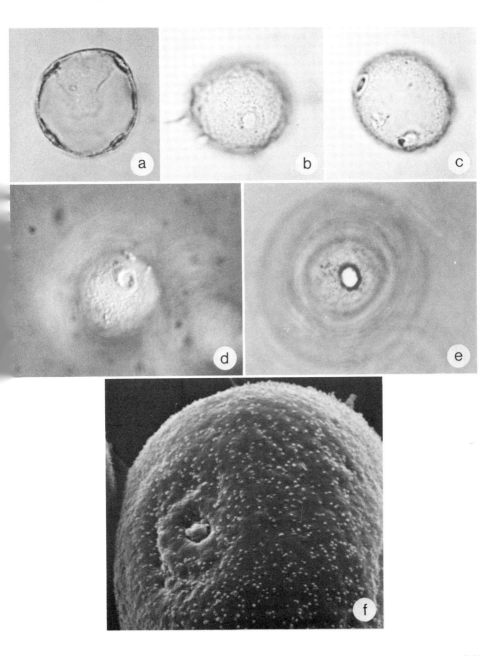

Description. Grains triporate or tetraporate (ratio 1:1); in polar view circular, 26–32 μm, av 29 μm in diam; in meridional view elliptical, 24–31 μm, av 27 μm in width; pores slightly aspidate, annulate, operculum present on unacetolyzed material, circular, 2–3 μm in diam, annulus prominent, 3.5 μm in diam; sculpturing scabrate rugulate; wall 1 μm thick; structure tectate.

SEM: Sculpturing, microscabrate particles unevenly distributed over the entire grain surface.

Flowering. April and May.

Native to. Eastern North America.

Distribution. Extreme southern part of Manitoba, southern Ontario and Quebec; not abundant.

Notes. According to Wodehouse (1971) the importance of the hackberries in causing hay fever is not well understood.

Pollen Key to the Urticaceae in Canada

A. Grains diporate, rarely triporate

    B. Grains with microechinate sculpturing (mucronate tip), spine unevenly spaced under the SEM .... **Boehmeria cylindrica**

    B. Grains without microechinate sculpturing, elements unevenly spaced under the SEM

        C. Grains with microtuberculate sculpturing under the SEM .... **Pilea pumila**

        C. Grains with a flattened top, elevated cylindrical sculpturing under the SEM .... **Laportea canadensis**

A. Grains triporate or tetraporate, rarely diporate

    D. Pores with a large distinct annulus, microechinate sculpturing under the SEM .... **Urtica dioica**
                                    U. dioica ssp. **gracilis**
                                         U. urens

    D. Pores with a small annulus, microechinate sculpturing under the SEM .... **Parietaria pensylvanica**

URTICACEAE

Boehmeria cylindrica (L.) Sw.   False nettle, bog-hemp.

Description.   Grains diporate, rarely triporate, spheroidal or occasionally ovoid when viewed with both pores in equatorial optical section, 12.5–13.5 μm, av 13.0 μm in diam; pore margin with a slight annulate thickening and pore occasionally operculate; sculpturing microechinate, small distinct granules of various sizes; ektexine and endexine 1 μm thick; structure intectate.

SEM: Sculpturing mucronate-echinate with spines unevenly spaced.

Flowering.   Mid-July to early August.

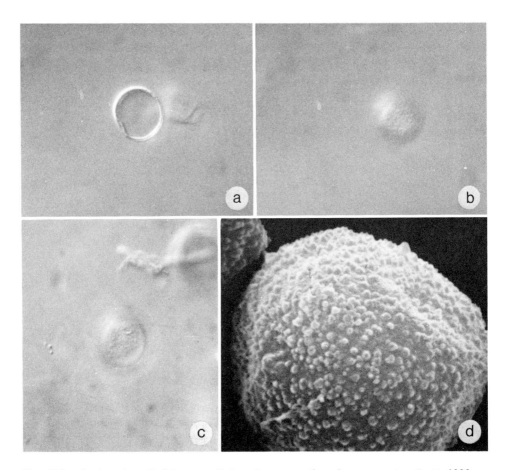

Fig. 136. *Boehmeria cylindrica. a–c,* light microscope (interference contrast), × 1000, *a,* equatorial section, *b,* surface, *c,* surface and pore; *d,* SEM, × 8250.

Native to. North America.

Distribution. Southern Ontario to southeastern Quebec.

Notes. When shed in large amounts, the pollen may cause hay fever.

URTICACEAE

Laportea canadensis (L.) Wedd. Wood nettle, nettle.

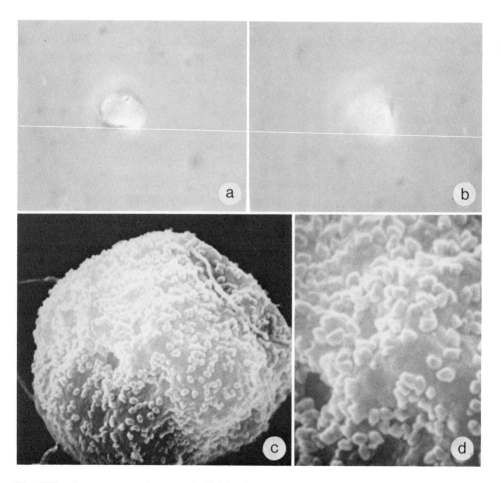

Fig. 137. *Laportea canadensis. a–b*, light microscope (interference contrast), × 1000, *a*, surface and pore, *b*, surface; *c–d*, SEM, *c*, × 8900, *d*, × 19 250.

Description. Grains diporate, rarely triporate; in polar and meridional views circular, 11.0–13.0 μm, av 12.0 μm in diam; pore 1.5 μm in diam with no distinct annulus, operculum present; sculpturing, unevenly spaced elements; ektexine and endexine about 1 μm thick; structure intectate.

SEM: Sculpturing flattened, top-elevated particles unevenly spaced over the surface.

Flowering. Early July to September.

Native to. North America.

Distribution. Western Saskatchewan to Newfoundland.

Notes. Pollen shed from this species in large amounts may cause hay fever.

URTICACEAE

Parietaria pensylvanica Muhl.   Pellitory

Description. Grains triporate, occasionally tetraporate; in polar and meridional views mostly circular, 11.5–16.5 μm, av 15 μm in diam; pore 1.5 μm in diam, small annulus, operculum present; sculpturing microechinate; ektexine and endexine about 1 μm thick; structure intectate.

SEM: Sculpturing microechinate, acute evenly spaced.

Flowering. Mid-June to the end of August.

Native to. North America.

Distribution. British Columbia to southeastern Quebec.

Notes. Because this species is not common in any given area, it is doubtful if enough pollen is shed to cause hay fever.

(Fig. 138 overleaf)

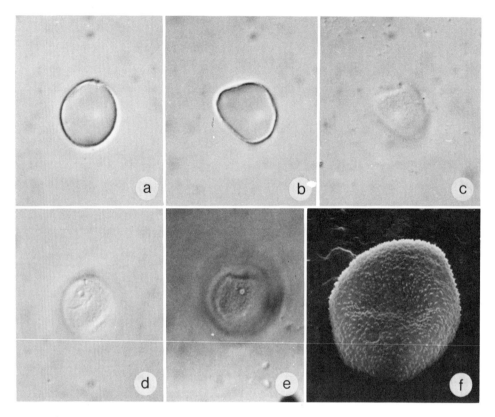

Fig. 138. *Parietaria pensylvanica.* a–e, light microscope (interference contrast), × 1000, a–b, equatorial section, c, surface, d, surface and pore, e, surface, pore, and annulus; f, SEM, × 3650.

URTICACEAE

Pilea pumila (L.) Gray Richweed.

Description. Grains diporate, rarely triporate; in polar and meridional views mostly circular, 13.5–15.5 μm, av 14.0 μm in diam; pore 1.5–2.0 μm in diam, annulus and operculum present; sculpturing, unevenly spaced particles; ektexine and endexine about 1 μm thick; structure intectate.

SEM: Sculpturing microtuberculate, elements unevenly spaced over the surface.

Flowering. Mid-July to early October.

Native to. North America.

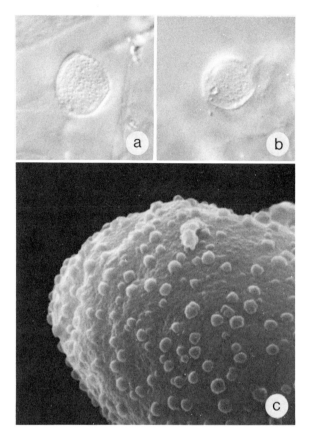

Fig. 139. *Pilea pumila.* a–b, light microscope (interference contrast), × 1000. a, surface, b, surface and pore; c, SEM, × 7000.

Distribution. Southeastern Ontario to Prince Edward Island.

Notes. When shed in large amounts, the pollen of richweed may cause hay fever.

URTICACEAE

Urtica dioica L. ssp. *gracilis* (Ait.) Selander. Stinging nettle, burning nettle, nettle.

Description. Grains triporate, occasionally tetraporate, rarely diporate; in polar and meridional views mostly circular, 11–18 μm, av 15 μm in diam; pore 2–3 μm in diam, annulus and operculum present; sculpturing microechinate; ektexine and endexine 1 μm thick; structure intectate.

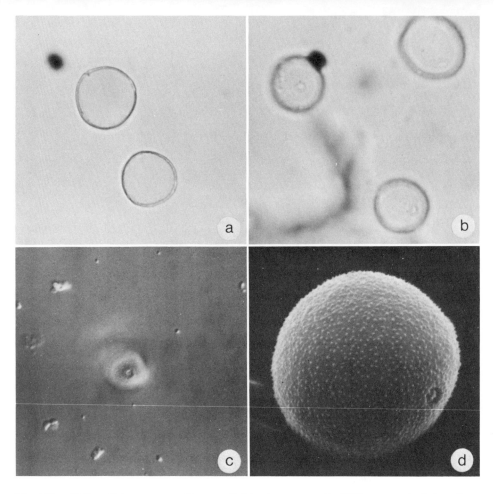

Fig. 140. *Urtica dioica* ssp. *gracilis*. *a–c*, light microscope, × 1000, *a*, equatorial section, *b*, surface and pore, *c*, surface under interference contrast; *d*, SEM, × 3700.

SEM: Sculpturing microechinate; pores with a large distinct annulus.

Flowering. Late May to October.

Native to. North America.

Distribution. British Columbia to Newfoundland and in the Yukon and Northwest Territories.

Notes. Hyde (1959) using a volumetric spore trap at Cardiff, Wales, made an assessment of the numerical abundance of airborne urticaceous pollen whose relative importance by reason of small size had previously been underestimated. The amount of nettle pollen approached that of grass pollen. The pollen of stinging nettle may play a bigger role in causing hay fever than grass pollen.

GLOSSARY

In general the descriptive terms used in this atlas are contained in Morphologic Encyclopedia of Palynology (Kremp 1965).

aciculate   Marked with fine irregular streaks.
annulus (annulate adj.)   A ringlike thickening of the ektexine around a pore.
arcus   Bandlike concave thickenings of the ektexine that delimit the pores, particularly in *Alnus* pollen.
aspidate   Pores protruding and borne on more or less circular areas or rounded domes.
baculate   Structural elements that are rodlike when viewed in profile.
colpi   Germinal furrows located in the meridional zone of the grain that usually extend at right angles to the equator toward the poles (excepting *Rumex*).
colporate   With furrows and pores.
columellae   Small rodlike, pillar structures connecting the tectum to the ektexine.
dicolporate   Possessing two furrows and two pores.
diploxylon   A type of pine pollen with bladders that are semicircular in polar view and broadly attached to the body of the grain. The furrow is not bounded by a thickened rim and its limits are not clearly defined.
diporate   With two pores.
disaccate   Possessing two wings, bladders, or air sacs.
distal   The part directed away from the tetrad connection.
dorsal cap   In gymnospermous pollen the exine surface opposite the germinal colpi.
echinate   Having spiny sculptural elements more than 1 μm high.
ektexine   The outer of the two layers of exine.
elaters   Hydroscopic bands or straps with flattened ends that encircle unacetolyzed *Equisetum* spores.
elliptic   Shaped like an ellipse, oblong with rounded ends.
endexine   The distinct inner layer of exine beneath the ektexine.
equator   The area encircled by an imaginary line midway between the poles of pollen grains.
equatorial axis   The axis perpendicular to the main axis, which is a line connecting the poles.
exine   The outer layer of the pollen wall.
extraporate   With two or more distinct apertures, which also possess lacunae.
fissura   A crack or opening usually found on inaperturate pollen.
foveolate   A sculpturing type consisting of minute holes or pits 1 μm or less in diameter.
furrow   *See* colpi.
gemmate   Having a flattened granule with the base constricted so that the granule is larger in diameter than in height.
glomerate   A sculpturing type comprised of clustered irregular particles.

haploxylon A type of pine pollen with bladders and grain body forming three intersecting circles in polar view; the furrow is usually clearly defined, has distinct limits, and is bordered by thick granular verrucate elements.
heteropolar With subisopolar grains possessing pores where there are morphological or shape differences between the distal and proximal faces.
inaperturate Having no pores, furrows, or visible openings.
intectate Describes grains upon which the ektexine sculpturing elements (if present) are free and isolated from each other.
intercolpium The ektexine contained or positioned between the colpi or margo on bipolar pollen grains.
intine The innermost layer of the pollen wall, which is not very resistant to acetolyzation and is usually destroyed.
isopolar The grains have no differences between the hemispheres or the proximal and distal faces.
lacuna(e) A large pore opening; the lacunae are separated by thickened ridges.
lumina The spaces or meshes of a reticulum that are bordered by the muri.
marginal ridge The proximal crest found in pollen of some types of Coniferae.
micro A prefix used to indicate that the various sculpturing type elements are less than 1 μm in size.
monoaperturate With one pore or furrow.
monocolpate With one germinal furrow.
monoporate With one pore.
muri The ridgelike network that forms the border of the meshes to a reticulum.
nexine The inner nonsculptured part of the exine, which corresponds to the endexine.
nonpendulous Describes bladders that are not positioned directly beneath the main body of saccate pollen.
oblate Distinctly flattened at the poles with the equatorial width wider than the length from pole to pole.
operculate Having an ektexinuous plug or cap over a pore or furrow.
pentacolpate Having five furrows.
pericolporate With pores and furrows evenly distributed over the surface.
periporate With several pores uniformly distributed over the grain surface but not equally along the equator (including subisopolar grains in Juglandaceae).
persorption The process by which particles as large as pollen and spores pass directly from the stomach or intestinal lumen epithelial cells into the blood of dogs and man.
pila Rodlike structures or clavate processes positioned closely together so that the heads nearly touch and give the impression of a network.
pilate With small rods (pila).
polar area The area of the pole on bipolar grains delimited by the ends of the colpi.

polar area index The ratio between the polar area dimension (space between colpi at the pole in polar view) and the largest transverse equatorial dimension of the grain (in polar view).
prolate Having isopolar grains in the shape of an ellipse with extended or elongated poles; the length of the polar axis is larger than that of the equatorial diameter.
protuberance A prominent bulging out.
proximal The inward part where tetrads are connected.
psilate Sculpturing absent or nearly so.
reticulate The sculpturing elements form a network.
reticulum Grains with muri and lumina, where the lumina are larger than the muri.
rugulate A sculpturing type comprised of elongated irregular elements.
scabrate With small sculpturing projections less than 1 μm in height (usually appearing to be microechinate under the SEM).
sculpture Surface appearance without reference to the structures.
sexine The outer sculptured part of the exine (*see* ektexine).
sinus The recessed inward curve between two projections (used to describe the germinal area contained between the bladders on some saccate pollen types).
stephanocolporate A grain type with four or more meridional furrows, each having an equatorial pore.
stephanoporate Having four or more meridional pores.
striate A type of sculpturing where the elements are streaked in more or less parallel bands or grooves.
structure The wall of the grain when viewed in profile, sectioned optically or mechanically.
subisopolar Where slight morphological features or shape differs from one hemisphere or pole to another, or slight differences occur between the proximal and distal faces.
suboblate With a shape where the ratio between the polar and equatorial axes is 0.75–0.88.
tectate Applies to grains that possess a tectum.
tectum The formation of a second rooflike membrane by the ektexine whether separated by a distinct cavity or not.
tetracolpate With four furrows.
tetracolporate With four furrows and four pores.
tetrad Pollen grains or spores grouped in fours.
tetragonal Four-sided shape.
tetraporate With four pores.
tricolpate Possessing three colpi or furrows.
tricolporate Having three colpi or furrows containing three pores.
trilete A pollen grain or fern spore possessing a triradiate tetrad scar.
triporate Having three pores.
triradiate A mark made on the proximal face of the exine by being attached or squeezed together as tetrads in the anther.
tuberculate A sculpturing type comprised of globose elements.
tubercules Tuberculate projections.

tumescense The swelling or enlargement of the ektexine, usually near an aperture or furrow.
undulating A wavy surface.
valla Ridges of striate or rugulate sculpturing.
verrucate Having warty irregularly shaped sculpture projections with the base usually twice the dimension of the height.
vestibulate Having a compartment (vestibule) between the outer and inner pore resulting from a separation of the ektexine and endexine.
vesiculate Possessing wings, bladders, or air sacs.

REFERENCES

Adams, R. J., and Morton, J. K. 1972. An atlas of pollen of the trees and shrubs of eastern Canada and the adjacent United States. Part I. Univ. Waterloo Biol. Series. No. 8. 52 pp.

Adams, R. J., and Morton, J. K. 1974. An atlas of pollen of the trees and shrubs of eastern Canada and the adjacent United States. Part II. Univ. Waterloo Biol. Series No. 9. 53 pp.

American Academy of Allergy. 1946. Report of the national pollen survey committee. J. Allergy 17:178.

American Academy of Allergy. 1947. Report of the national pollen survey committee. J. Allergy 18:284.

Bassett, I. J. 1956. Atmospheric pollen studies at Ottawa, Ontario. Can Dep. Agric. 21 pp.

Bassett, I. J. 1959. Surveys of airborne ragweed pollen in Canada with particular reference to sites in Ontario. Can. J. Plant Sci. 39:491–497.

Bassett, I. J. 1964. Airborne pollen surveys in Manitoba and Saskatchewan. Can. J. Plant Sci. 44:7–14.

Bassett, I. J. 1965. Pollen grains and plant history. Greenhouse-Garden-Grass. 5(2):5–6.

Bassett, I. J. 1973. The plantains of Canada. Can. Dep. Agric. Monogr. No. 7. 47 pp.

Bassett, I. J., and Crompton, C. W. 1966. Airborne pollen surveys in British Columbia. Can. J. Plant Sci. 47:251–261.

Bassett, I. J., and Crompton, C. W. 1968. Pollen morphology and chromosome numbers of the family Plantaginaceae in North America. Can. J. Bot. 46:349–361.

Bassett, I. J., and Crompton, C. W. 1969. Airborne pollen surveys in eastern Canada. Can. J. Plant Sci. 49:247–253.

Bassett, I. J., and Crompton, C. W. 1975. The biology of Canadian weeds. 11. *Ambrosia artemisiifolia* L. and *A. psilostachya* DC. Can J. Plant Sci. 55:463–476.

Bassett, I. J., Crompton, C. W., and Frankton, C. 1976. Canadian havens from hay fever. Can Dep. Agric. Publ. 1570. Ottawa. 23 pp.

Bassett, I. J., Holmes, R. M., and MacKay, K. H. 1961. Phenology of several plant species at Ottawa, Ontario, and an examination of the influence of air temperature. Can. J. Plant Sci. 41:643–653.

Bassett, I. J., Mulligan, G. A., and Frankton, C. 1962. Poverty weed, *Iva axillaris*, in Canada and the United States. Can. J. Bot. 40:1243–1249.

Bassett, I. J., and Terasmae, J. 1962. Ragweeds, *Ambrosia* species in Canada and their history in postglacial time. Can. J. Bot. 40:141–150.

Baum, B. R., Bassett, I. J., and Crompton, C. W. 1970. Pollen morphology and its relationship to taxonomy and distribution of *Tamarix*, series Vaginantes. Osterr. Bot. Z. 118:182–188.

Best, K. F. 1975. The biology of Canadian weeds. 10. *Iva axillaris* Pursh. Can. J. Plant Sci. 55:293–301.

Dugle, J. R., and Coops, T. P. 1972. Pollen characteristics of Manitoba cattails. Can. Field Nat. 86(1):33–40.

Durham, O. C. 1937. Evaluation of the ragweed hay fever resort areas of North America. J. Allergy 8:175.

Durham, O. C. 1951. The pollen harvest. Econ. Bot. 5:211–254.

Erdtman, G. 1957. Pollen and spore morphology. Plant taxonomy: Gymnospermae, Pteridophyta, Bryophyta. An introduction to palynology. II. Almqvist & Wiksell, Stockholm, Swed. 151 pp.

Erdtman, G. 1966. Pollen morphology and plant taxonomy. Angiosperms. Hafner Press, New York, N.Y. 553 pp.

Erdtman, G. 1969. Handbook of palynology. Hafner Press, New York, N.Y. 486 pp.

Erdtman, G., Berglund, B., and Praglowski, J. R. 1961. An introduction to a Scandinavian pollen flora. Grana Palynol. 2(3):3–92.

Erdtman, G., Praglowski, J. R., and Nilsson, S. 1963. An introduction to a Scandinavian pollen flora. II. Almqvist & Wiksell, Stockholm, Swed. 89 pp.

Faegri, K., and Iversen, J. 1964. Textbook of pollen analysis. 2nd rev. ed. Hafner Press, New York, N.Y. 237 pp.

Feinberg, S. M. 1946. Allergy in practice. Year Book Medical Pubs. Inc. Chicago, Ill. 838 pp.

Frankton, C., and Mulligan, G. A. 1970. Weeds of Canada. Can. Dep. Agric. Publ. 948. Queen's Printer, Ottawa, Ont. 217 pp.

Gebben, A. I. 1965. The ecology of common ragweed, *Ambrosia artemisiifolia* L., in southeastern Michigan. University Microfilms Inc., Ann Arbor, Mich. 234 pp.

Gleason, R. A., and Cronquist, A. 1963. Manual of vascular plants of northeastern United States and adjacent Canada. Van Nostrand, Princeton, N.J. 810 pp.

Godwin, H. 1956. History of the British flora. Cambridge University Press, London, England. 384 pp.

Hedberg, O. 1946. Pollen morphology in the genus *Polygonum* L. s. lat. and its taxonomic significance. Sven. Bot. Tidskr. 40:371–404.

Helmich, D. E. 1963. Pollen morphology in the maples (*Acer* spp.) Pap. Mich. Acad. Sci. Arts Lett. 48:151–164.

Heusser, C. J. 1971. Pollen and spores of Chile. University of Arizona Press, Tucson, Ariz. 167 pp.

Hirst, J. M. 1952. An automatic volumetric spore trap. Ann. Appl. Biol. 39:257–265.

Hitchcock, C. L., Cronquist, A., Ownbey, M., and Thompson, J. W. 1961. Vascular plants of the Pacific Northwest. Part 3. Saxifragaceae to Ericaceae. Univ. Wash. Publ. Biol. 17(3):1–614.

Hitchcock, C. L., Cronquist, A., Ownbey, M., and Thompson, J. W. 1964. Vascular plants of the Pacific Northwest. Part 2. Salicaceae to Saxifragaceae. Univ. Wash. Publ. Biol. 17(2):1–597.

Holmes, R. M., and Bassett, I. J. 1963. Effect of meteorological events on ragweed pollen count. Int. J. Biometeorol. 7:27–34.

Hosie, R. C. 1969. Native trees of Canada. 7th ed. Can. For. Serv. Dep. Fish. and For. Queen's Printer, Ottawa, Ont. 380 pp.

Hyde, H. A. 1959. Volumetric counts of pollen grains at Cardiff, 1954–57. J. Allergy 30:219–234.

Hyde, H. A. 1972. Atmospheric pollen and spores in relation to allergy. I. Clin. Allergy 2:153–179.

Hyde, H. A., and Adams, K. F. 1958. An atlas of airborne pollen grains. Macmillan & Co. Ltd., London, England. 112 pp.

Kapp, R. O. 1969. How to know pollen and spores. William C. Brown Co., Dubuque, Iowa. 249 pp.

Kennedy, L. L. 1953. Alberta pollen survey. J. Allergy 24:355–363.

Kremp, G. O. W. 1965. Morphologic encyclopedia of palynology. University of Arizona Press, Tucson, Ariz. 186 pp.

Kuprianova, L. A. 1956. The structure of the membrane of pollen grains. Bot. Zh. Akad, Nauk SSSR 41(8):1212–1216.

Larson, D. A., Skvarla, J. J., and Lewis, C. W. Jr. 1962. An electron microscope study of exine stratification and fine structure. Pollen et Spores 4(2):234–246.

Linskens, H. F., and Jorde, W. 1974. Persorption of *Lycopodium* spores and pollen grains. Die Naturivissenchaften 61:275–276.

Little, E. L. Jr. 1953. Check list of native and naturalized trees of the United States. Agric. Handbook No. 41. U.S. Government Printing Office, Washington, D.C. 472 pp.

Martin, P. S., and Drew, C. M. 1969. Scanning electron photomicrographs of southwestern pollen grains. J. Ariz. Acad. Sci. 5:147–176.

Martin, P. S., and Drew, C. M. 1970. Additional scanning electron micrographs of southwestern pollen grains. J. Ariz. Acad. Sci. 6:140–161.

McAndrews, J. H., Berti, A. A., and Norris, G. 1973. Key to the quarternary pollen and spores of the Great Lakes region. Life Sci. Misc. Publ. Royal Ontario Museum, Toronto, Ont. 61 pp.

Newmark, F. M., and Itkin, I. H. 1967. Asthma due to pine pollen. Ann. Allergy 25:251.

Ogden, E. C., Raynor, G. S., Hayes, J. V., Lewis, D. M., and Haines, J. H. 1974. Manual for sampling airborne pollen. Macmillan Publishing Co., Inc., New York, N.Y. 182 pp.

Ogden, E. G., Raynor, G. S., and Hayes, J. V. 1975. Travels of airborne pollen. Publ. no. EPA-650/3-75-003. Ecol. Res. Series. U.S. Environmental Protection Agency, Washington, D.C. 100 pp.

Ontario Ministry of Agriculture and Food. 1973. Agricultural statistics for Ontario. Publ. 20. 91 pp.

Payne, W. W. 1964. A re-evaluation of the genus *Ambrosia* (Compositae). J. Arnold Arbor. Harv. Univ. 45:401-438.

Pellett, F. C. 1947. American honey plants. 4th ed. Orange Judd Publishing Co., New York, N.Y. 467 pp.

Praglowski, J. R. 1962. Notes on the pollen morphology of Swedish trees and shrubs. Grana Palynol. 3(2):45-65.

Raynor, G. S., and Ogden, E. C. 1970. The swing-shield: an improved shielding device for the intermittent rotoslide sampler. J. Allergy 45:329-332.

Richard, P. 1970. Atlas pollinique des arbres et de quelques arbustes indigènes du Québec. Nat. Can. (Que.) 97(1):1-34; (2):97-161; (3):241-306.

Samter, M., and Durham, O. C. 1955. Regional allergy of the United States, Canada, Mexico, and Cuba. Charles C. Thomas, Publisher, Springfield, Ill. 395 pp.

Solomon, A. M., and Buell, M. F. 1969. Effects of suburbanization upon airborne pollen. Bull. Torrey Bot. Club 96(4):435-444.

Smith, R. D., and Rooks, R. 1954. The diurnal variation of airborne ragweed pollen as determined by a continuous recording particle sampler and implications of the study. J. Allergy 25:36-45.

Stenburg, R. L., and Hall, L. B. 1951. A continuous recording particle sampler. Proc. 5th meeting, Northeast Weed Control Conference, pp. 311-317.

Tai-June, Y., Spitz, E., and McGerity, J. L. 1974. Allergy to Cupressaceae pollen. J. Allergy Clin. Immunol. 5:71-72.

Van Campo-Duplan, M. 1954. Considerations generales sur les caractères des pollens et des spores et sur leur diagnose. Bull. Soc. Bot. Fr. 101(5-6):250-281.

Vaughan, W. T., and Black, J. H. 1948. Practice of allergy. 2nd ed. C. V. Mosby Co., St. Louis, Mo. 1132 pp.

Voisey, P. W., and Bassett, I. J. 1961. A new continuous pollen sampler. Can. J. Plant Sci. 41:849-853.

Whitehead, D. R. 1963. Pollen morphology in the Juglandaceae: I. Pollen size and pore number variation. J. Arnold Arbor. Harv. Univ. 44(1):101–111.

Whitehead, D. R. 1965. Pollen morphology in the Juglandaceae: II. Survey of the family. J. Arnold Arbor. Harv. Univ. 46(4):369–410.

Wodehouse, R. P. 1935. Pollen grains. McGraw-Hill Book Co., New York, N.Y. 574 pp.

Wodehouse, R. P. 1971. Hay fever plants. 2nd rev. ed. Hafner Press, New York, N.Y. 280 pp.

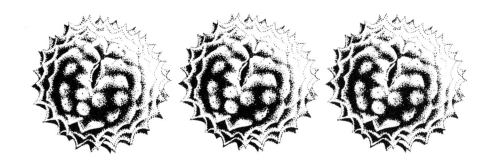

Part 2　COMMON AIRBORNE FUNGUS SPORES

Fungi abound in our environment and affect us in many ways. They cause diseases in humans, in domestic and wild animals, and in cultivated and wild plants. These organisms cause rots, decays, and molding in natural and processed animal and plant matter. They play a necessary role in the breakdown of dead organic matter into the elements required by green plants. Fungi do not possess chlorophyll and must obtain their nutrients from elaborated organic material. They may be microscopic in soil or on plants or animals and virtually undetected by anyone not familiar with them, or they may be large and easily seen as fleshy mushrooms or as conks on rotting wood. Many fleshy fungi are edible (*Agaricus, Morchella*), but some are poisonous if eaten (*Amanita, Gyromitra*). Some microfungi produce chemicals that are poisonous if consumed with other foods (*Aspergillus*, aflatoxin; *Claviceps*, ergot), and some are beneficial in combating bacterial infections (*Penicillium*, penicillin).

The fungi are classified on the characters of their vegetative bodies, sporogenous bodies, and spores (Table 2). Allergenic fungi are known or suspected in all classes but especially among the higher fungi.

The higher fungi possess a vegetative body of cells joined end to end into long slender hyphae, which in mass is called mycelium. Certain hyphae (conidiophores) take on a special role to form the reproductive units called spores. These sporogenous hyphae may appear as aerial conidiophores directly on the mycelium (*Oidiodendron, Curvularia*), or they may form a mycelial mass with characteristic structures (uredinia of *Puccinia*, pycnidia of *Septoria*). These spores are asexual and their nuclear condition is exactly that of the parent mycelium. Sexual spores having half the chromosome number (haploid) of the parent develop in characteristic structures also (basidiocarps of *Agaricus*, ascocarps or cleistothecia of *Erysiphe*). Either as a spore or mycelium, the sexual state must fuse with a compatible component to reestablish the parental (diploid) chromosome number. Either kind of spore may be aerially disseminated, but some of either kind may be spread by water, insects, or animals. It is the airborne spores (Fig. 141) that are of concern as allergens.

Ontogeny, in particular for Fungi Imperfecti, has received renewed attention since time-lapse photography and electron microscopy have been developed. The ontogeny of *Oidiodendron* is shown in Fig. 142; a spore forms at the apex of the sporogenous hypha and then successive spores form toward the base. Powdery mildews (e.g. *Erysiphe*) also produce conidia in this way. Spores may also form from the base to the apex, and there are still other ways. Kendrick (1971) has edited the report of a workshop conference dealing in part with spore ontogeny in the Fungi Imperfecti.

Table 2. Classification of the fungi

Fungus class	Vegetative body	Sporogenous body	Spores
Slime molds			
Myxomycetes	plasmodium	sporangium	resting spores, motile zoospores
Lower fungi			
Phycomycetes	plasmodium or mycelium aseptate	sporangium or conidiophore	motile zoospores or conidia
Higher fungi			
Ascomycetes	mycelium septate	ascus	ascospores
Basidiomycetes	mycelium septate	basidium	basidiospores
Fungi Imperfecti*	mycelium septate	conidiophore	conidia

*These are asexual states of fungi some of which have been connected to the sexual states in Ascomycetes and Basidiomycetes, but many are still unconnected.

Fig. 141. *Lycoperdon pyriforme* illustrating spore release in mass. (Photo courtesy the late H. S. Cooke).

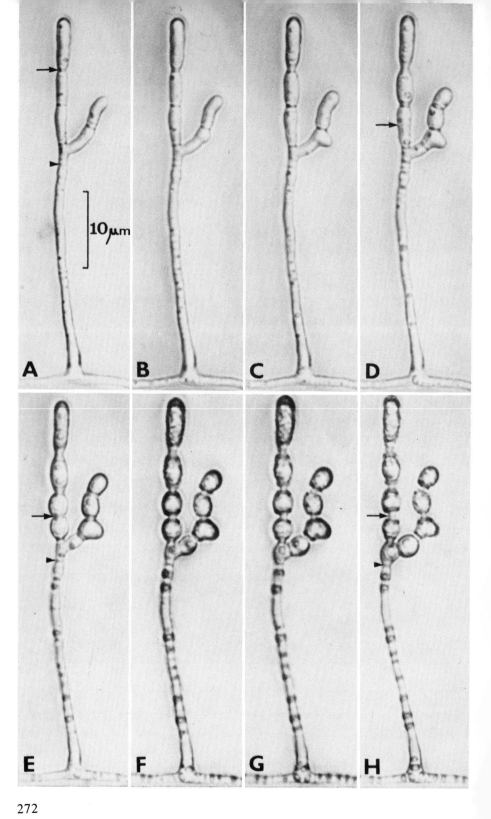

Many spores (e.g. conidia of Fungi Imperfecti or rust urediniospores) are so delicately attached that slight air movements or any disturbance of the substrate causes them to become airborne. Other spores (ascospores or basidiospores) are forcibly discharged into the surrounding air.

Once airborne, spores are disseminated locally or often lifted as spore clouds. *Cladosporium* has been recorded in such concentrations over the North Sea as they were borne on westerly winds from England toward Denmark. The same study disclosed that whereas *Cladosporium* concentrations were carried at an altitude of 0.5-1.5 km, heavier pollen grains were borne in a similar pattern but at somewhat lower altitudes. Studies of rust epidemiology in North America show that urediniospores retain viability after aerial dissemination of at least 805 km from the Mississippi Valley northward to the Canadian prairies. This aspect of spores in the air has been summarized by Ingold (1971). In the same work he treats the release of spores under the chapter "Periodicity."

Maximum spore release for *Cladosporium* takes place in the summer (Ingold 1971), whereas ascospores of *Hypoxylon* peak in the autumn. Hyde (1972) has presented a pollen and spore calendar for Cardiff, Wales. In contrast with pollen, which occurs mainly in the first half of the year, fungus spores occur all year but mainly in the spring, summer, and autumn. *Cladosporium*, for instance, occurs all year and peaks in June through September. Spores of the fungi trapped vary from region to region. Records of the 24-h spore deposit on slides exposed in Paris, France, and at Pittsburgh, Pa., are presented in the graphs in Fig. 143. The data were compiled from the statistical report for 1970 of the pollen and weed committee of the American Academy of Allergy. Readings reflect the outdoor spore load; although peaking is obvious, spores are present throughout the year. No significance should be inferred from the relative abundance of spores between centers or to the presence or absence of species. In Canada much of the allergic trouble with children is in spring because children playing on lawns stir up the spores being formed on old leaves.

Superimposed on seasonal variation is a daily variation. Aeciospore discharge of *Puccinia andropogonis* and *Uromyces psoraleae* and basidiospore discharge of *Puccinia malvacearum* is at a maximum during the night; whereas urediniospores of *Puccinia recondita*, chlamydospores of *Ustilago nuda*, and conidia of *Cladosporium* build up to a peak around midday. Different fungi require specific periods of light and darkness, relative humidity, and temperature to produce spores and they depend on these factors plus other external ones, such as wind velocity, for release of these spores. There is a relationship between factors governing spore production and release for some taxonomic

---

142. *Oidiodendron truncatum*, time-lapse sequence of conidium formation. sequence is irregularly basipetal in 69½ h. The arrowhead is a common erence point and the arrows indicate transverse septa. (Photo courtesy W. B. ndrick).

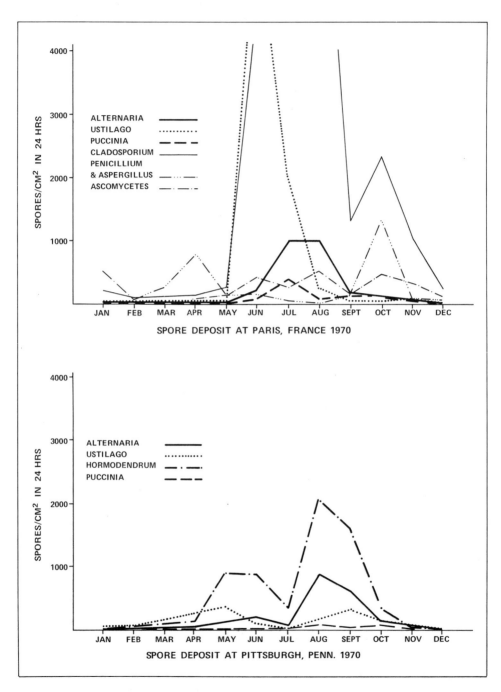

Fig. 143. Spore deposit graphs.

groups of fungi and for some ecologically related fungi but not for all fungi. These factors have been considered in a comprehensive text on aerial microbiology by Gregory (1961).

In the following pages examples of known or potentially allergenic fungi are described and illustrated. At the end of the chapter there are five plates of fungus spores arranged taxonomically for purposes of comparison.

DESCRIPTIONS OF FUNGUS SPORES

Alternaria alternata (Fr.) Keissler   (*Alternia tenuis* Nees)

Fungi Imperfecti   Hyphomycetes

Fructification.   Brown, septate mycelium gives rise to aerial septate, simple, solitary or fascicled conidiophores that may produce a single apical spore or a number of spores sympodially.

Spores.   Solitary or catenate, pale brown to green brown, ovoid or obclavate often with a prolongation at the paler distal end, horizontally and obliquely to vertically septate, 20–60 × 9–18 μm; wall minutely verrucose.

Spore release.   Spores are released throughout the year by the breaking apart of the spore chains.

Substrate and distribution.   It is found on living plants, plant matter including textiles, and many other substrates. The distribution is worldwide but is rare or lacking in arctic and alpine regions.

Notes.   Many species of *Alternaria* have been described; some are parasitic on higher plants and others are saprophytic on organic matter. Because spores have a characteristic shape, color, and size, they are readily recognizable on slides or plates and are recorded in most aerobiological surveys. The spores are large for fungus spores but are capable of long-distance dispersal. Christensen (1965) estimated that spores of *Alternaria* may be dispersed close to 4800 km horizontally from an altitude of 1.6 km. *Alternaria* spores have been reported in abundance in samplings over wheat fields in India (Mishra and Scrivasta 1972), over prairie environments (Long and Kramer 1972; Walton and Dudley 1946), over city environments (Chaterjee and Hargreave 1973), and in the home environment (Lumpkins et al. 1973).

In Ontario *Alternaria* spores were in greatest abundance in the air during July and September in the area around London and St. Thomas (Barlow 1963). In the Toronto area about 10 years later, the greatest abundance was from June to October with additional peaks in January and February (Collins-Williams et al. 1973). The winter peaks probably resulted from long-distance spore dispersal. Pulmonary sensitivity in workers engaged in stripping bark

Fig. 144. *Alternaria alternata*. a, habit on *Allium cepa* L. (onion) dead leaves, × 1; b, conidia horizontally and vertically septate, scale line = 10 μm; c, conidia shallowly verrucose-rugose, SEM × 4000.

from maple logs has been attributed to the inhalation of spores of *Cryptostroma corticale* (Hyphomycetes) formed under the bark. However, Schlueter et al. (1973) have shown *Alternaria* to be responsible for a similar disorder in two workers employed in the wood-pulp industry. Both patients reacted positively when subjected to the inhalation challenge of a saline solution of *Alternaria* and to intracutaneous tests with the same fungus. Collins-Williams et al. (1973) reported that *Alternaria* and *Helminthosporium* gave the highest positive bronchial provocation reaction of the 11 molds tested.

Aspergillus flavus Link

Fungi Imperfecti   Hyphomycetes

Fructification.   Erect, rough, aerial conidiophores end in a globoid vesicle that supports one or two rows of metulae and phialides. Conidial heads are yellow to yellow green and columnar.

Spores.   Globoid, single-celled, dry, catenate, 3.0–6.0 μm in diam, mostly 3.5–4.5 μm (related species up to 10 μm in diam); wall shallowly echinulate-verrucose.

Spore release.   Spore chains readily break apart.

Substrate and distribution.   The spores may be isolated from soils, decaying vegetation, and stored seeds around the world.

Notes.   *A. flavus* and related species are even more widely used in industry than *A. niger*. The enzyme amylase is used in the production of industrial alcohols from starches. Proteolase is used in the production of soy sauce and the treatment of hides and silk. Lipolase and esterase are involved in the breakdown of fatty or waxy substances.

Many fungi are known to produce substances poisonous to animals including man. Aflatoxin is produced by *A. flavus* growing on seeds, especially those of high oil content. Poultry flocks have been affected by aflatoxin in feed containing contaminated peanut oil meal. The toxin chemistry has been reviewed by Arrhenius (1973).

Aspergillus niger van Tiegh.

Fungi Imperfecti   Hyphomycetes

Fructification.   Smooth aerial conidiophores terminate in a globoid vesicle bearing spore-producing cells or phialides on special cells called metulae. These conidial heads are blackish and often split into conspicuous columns of conidial chains.

Spores.   Globoid, single-celled, dry, catenate, readily coming apart, 3.0–5.0 μm in diam, (related species up to 10.0 μm in diam); wall prominently and coarsely verrucose under SEM, minutely so under the light microscope.

Spore release.   Chains of cells are dry and readily break up when mature to become airborne.

Substrate and distribution.   This is probably the most abundant species of *Aspergillus* and occurs in soil, on plants, and any organic substrate.

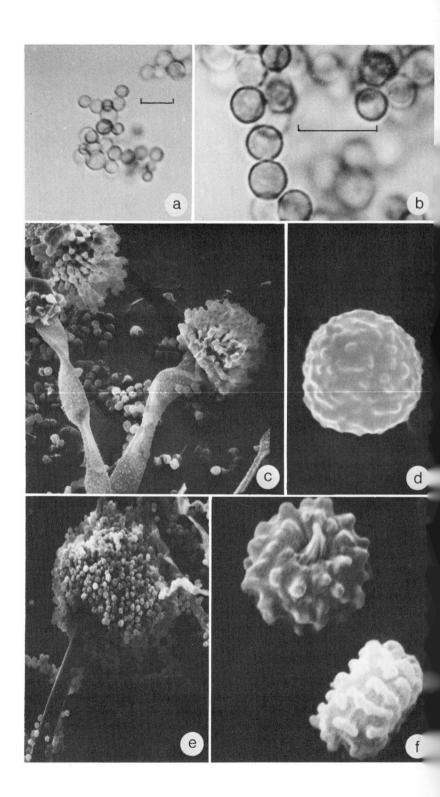

Notes. Spores of *Aspergillus* spp. are usually found in moderate counts in aerial surveys of pollen and spores. Recent reports of the occurrence of this mold in the air over Canada include: Chaterjee and Hargreave (1973) for Hamilton, Ont., Collins-Williams et al. (1973) for Toronto, Ont., and Barlow (1963) for southern Ontario.

A. niger and species in this complex have been reported as pathogenic in man and animals, but Austwick (1965) thinks there is little evidence to support the claims. He had little information regarding toxin production and his comments about allergy and antigenicity dealt mostly with *A. fumigatus* (a green-pigmented species).

The black aspergilli have been used in industry to produce citric, gluconic, and oxalic acids and amylolytic, lipolytic, pectolytic, and proteolytic enzymes. They have also been used to make pigments and antibiotics and in soil-testing procedures to determine mineral deficiencies (Raper and Fennel 1965).

Aureobasidium pullulans (de Bary) Arnaud (*Pullularia pullulans* [de Bary] Berkhout)

Fungi Imperfecti   Hyphomycetes

Fructification. Spores are formed directly on septate mycelium, which grows on various substrates from soil to milled wood, live plant parts including flower nectar, and decomposing plants.

Spores. Hyaline, ovate, slightly attenuate to point of attachment or hilum, 7–8 × 3–4 μm, tending to be smaller on young mycelium and larger on older mycelium; wall smooth, individual mycelial cells thickening into chlamydosporelike propagules (Cooke 1959).

Spore release. Airborne spores may become very abundant whenever the substrate is depleted of nutrients. Therefore no set time or period can be predicted because release correlates more with the substrate than with the season or climate.

Substrate and distribution. This fungus is found in temperate and tropical climates. In Europe it is considered to be a weak parasite on leaves; in America it is a storage rot organism of citrus and pomaceous fruits. In Australia it plays a role in the dewretting of fiber flax. The fungus grows on flower nectar

---

Fig. 145. *a–d*, *Aspergillus flavus*. *a–b*, conidia, scale lines = 10 μm; *c*, conidiophores, SEM × 735; *d*, conidium, SEM × 8400. *e–f*, *A. niger*, *e*, conidiophore, SEM × 735; *f*, conidia, SEM × 8330.

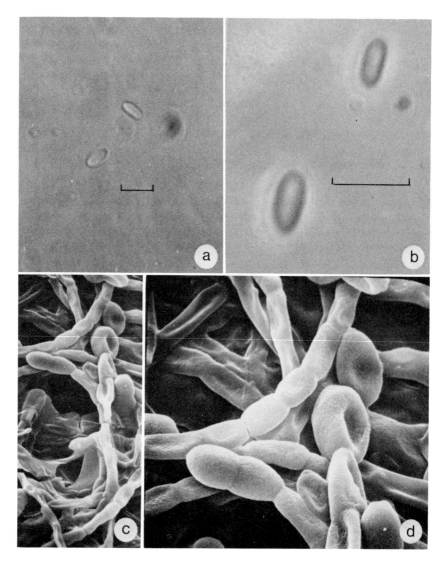

Fig. 146. *Aureobasidium pullulans. a–b*, conida, scale lines = 10 μm; *c–d*, conidia and mycelium, *c*, SEM × 735, *d*, SEM × 1470.

and is regularly reported as an isolate from seed. It is regularly found in varying amounts on slides exposed for airborne counts. Cooke (1959) reported that it is difficult to recognize this fungus from air-trapped spores, and he depends on observations of culture-produced spores for identifications. Widden and Parkinson (1973) found this fungus commonly in the litter of pine forests in Ontario and Alberta.

Notes. Cooke (1959) remarked that the fungus has not been shown to be important in allergies. However, Collins-Williams et al. (1973) in provocative bronchial testing with molds found positive reactions for *Aureobasidium* (*Pullularia*), which were less severe than for *Helminthosporium, Alternaria, Aspergillus, Hormodendrum* (see *Cladosporium*), or *Penicillium.*

Aureobasidium was the second predominant genus following *Cladosporium* in a culture plate survey of the airborne fungi of five widely separated and climatologically diverse sites in the United States and Puerto Rico (Sorenson et al. 1974). The three main fungi isolated in this study (the above plus *Penicillium*) were found consistently at all sites. Other fungi did not occur with the same consistency. *Monilia* was found rarely at all sites except at San Juan, Puerto Rico, where it was of relatively high occurrence. *Alternaria* was found rarely at San Juan and Tacoma but occurred in moderate abundance at all other sites.

Aureobasidium followed *Alternaria* as the most abundant airborne fungus trapped in lumber storage locations in the Ottawa Valley in 1971 (Unligil et al. 1974). Other fungi occurring abundantly were *Penicillium, Epicoccum, Nigrospora,* and *Cladosporium.* Plates exposed in the afternoon generally gave higher colony counts than those used in the morning.

Cladosporium herbarum (Pers.) Link ex S.F. Gray

Fungi Imperfecti   Hyphomycetes

Fructification. Aerial conidiophores arise from green brown mycelium, which may clump into a compressed mass called a stroma. The conidiophores are straight to flexuous, sometimes geniculate, and sometimes have nodose vescicular swellings. They are pale brown to brown or green brown, smooth, sometimes branched toward the apex, and generally ca. 25–100 μm long. Occasionally they are 175 μm long.

Spores. Conidia formed on the paler upper length of the conidiophore in long, sometimes branched, chains; ellipsoidal to oblong with a protruding hilum usually at both ends, 0–1 septate, occasionally with additional horizontal septa; 5–28 (–33) × 3–8 (–10) μm; wall pale to olivaceous brown, smooth to finely verrucose.

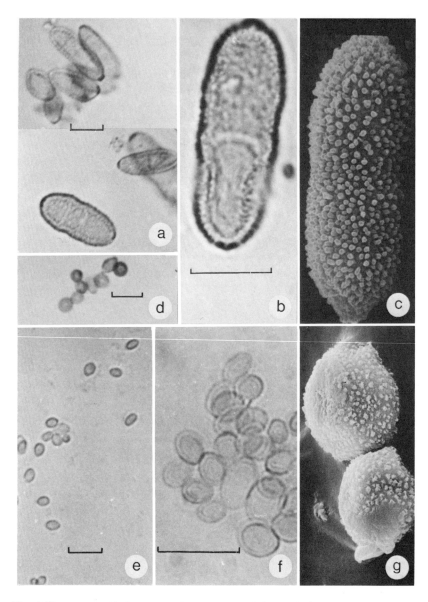

Fig. 147 *a–c, g, **Cladosporium herbarum**,* conidia; *c*, SEM × 4000, *g*, SEM × 4000. *d, C. sphaerospermum*, conidia. *e–f, **Hormodendrum*** sp. Scale lines = 10 μm.

Spore release. Pady et al. (1969) in periodicity experiments showed that *Cladosporium* spores are formed under cool and moist conditions at night. Spore release takes place during the day when the temperature rises and the relative humidity falls. In a wheat field in Kansas, sporulation peaked at 7-8 a.m. Morning turbulence and dropping relative humidity were considered important in spore release.

Substrate and distribution. This species has worldwide distribution and is common in temperate and arctic regions as a saprophyte of dead herbaceous plants, especially grasses, but also of woody plants, soil, foods, and textiles. Spores are regularly isolated from the air (Pady et al. 1969) and litter and humus in pine forests (Widden and Parkinson 1973).

Notes. Ellis (1971) illustrated and described 15 common species of *Cladosporium* but noted that some 500 species have been described. Some of these are parasitic on foliage of higher plants and cause well-defined leaf spots. Others are the conidial states of such Pyrenomycetes as *Mycosphaerella*, *Venturia*, and *Amorphotheca*. *C. sphaerospernum* has spores 3.0-4.5 µm in diameter and occurs commonly indoors on damp walls. In their study of woodland and prairie sites in Kansas, Long and Kramer (1972) found that *Cladosporium*, which is common on prairie grasses, was most abundant in their traps on the prairie during dry daytime weather. At the woodland sites *Cladosporium* spores were constant but at a low level throughout 24 h. They concluded that spores trapped at the woodland site were of local origin, whereas those trapped at the prairie site were partly from local and partly from a long-distance aerial source. *Cladosporium* made up 82% of the total spores trapped during air flights between Montreal and London according to Pady and Kapica (1955). They concluded that spore numbers were correlated with air masses and that spores could successfully cross the Atlantic Ocean.

The taxonomy of *Cladosporium* needs clarification. Hughes (1958) lists as synonyms *Myxocladium* Cda., *Didymotrichum* Bon., *Sporocladium* Chev., and *Heterosporium* Klotzsch. This concept has been followed by Ellis (1971) and Barron (1968), who stated that *Hormodendrum* is also generally considered as a synonym of *Cladosporium*. It is possible that *Hormodendrum* is only a cultural phase of *Cladosporium*,* but more cultural work is required to test this concept for the genus as a whole. Ainsworth and Bisby (1971) treat *Hormodendrum* as a synonym of *Cladosporium*.

Most surveys for airborne fungus spores report *Cladosporium* and *Hormodendrum*, separately or as combined taxa, as the most abundant of the fungi trapped. *Hormodendrum* was the main fungus trapped in the home environment according to Lumpkins et al. (1973). They found it throughout the year but mainly from June to November. During this period the incidence of these spores was greater indoors than outdoors; the reverse was true for

---

*Personal communication from S. J. Hughes.

fungi other than *Hormodendrum*. Following *Hormodendrum* in abundance were *Penicillium, Alternaria, Aspergillus,* and ? *Fomes*; altogether over 70 fungi were found. The spores were trapped on culture plates. In Toronto, Ont., Collins-Williams et al. (1973) used microscope slide traps and found that *Hormodendrum* was the predominant outdoor spore form and occurred most abundantly from late May to September. In the order of the numbers of spores found, they reported the commonest molds as *Hormodendrum, Alternaria,* smut, and yeast.

The fructifications and spores of *Hormodendrum* are similar to those of *Cladosporium*. The spores are usually nonseptate with pale coloring and at the smaller end of the size range; walls appear smooth under the light microscope.

Curvularia spp.

Fungi Imperfecti   Hyphomycetes

Fructification.   Spores are formed sympodially on brown, septate aerial conidiophores. They form apically and laterally and occur singly or in headlike clumps.

Spores.   Septate, pale to dark brown, darkest in the larger central cells, straight to curved, ellipsoid, broadly fusiform or obovoid, ends rounded, base sometimes having a protruding hilum, occasionally with twin apical cells; wall usually smooth, but variously verrucose in some species, thickest in central cells; spore size variable, e.g. *C. andropogonis,* 45–66 × 18–28 μm, 3-septate; *C. geniculata,* 28–36 (–42) × 11–14 μm, 4-septate; *C. trifolii,* 25–35 (–38) × 11–15 μm, 3-septate.

Spore release.   The spores are dry and detach readily from the conidiophore.

Distribution.   In temperate and tropical regions the genus is found worldwide; certain species are limited, presumably because of host restrictions.

Notes.   As plant parasites *Curvularia* spp. cause leaf spots and corm and root rots especially of grasses; as saprophytes they occur on many substrates especially in tropical regions.

In many air samplings, spores of *Curvularia* are reported in low to moderate abundance.

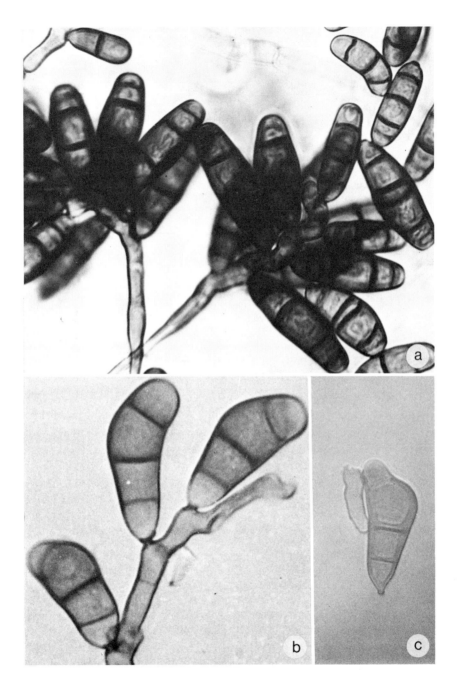

Fig. 148. *Curvularia* spp. *a*, *C. geniculata* conidia formed sympodially, × 900. *b*, *C. trifolii* f. sp. *gladioli*, × 1300. *c*, *C. trifolii* showing protruding hilum at base of conidium, × 900.

Curvularia spp. were reported among spores trapped over wheat fields in India (Mishra and Srivastava 1972) and among spores trapped in a survey of homes (Lumpkins et al. 1973). In a survey of the fungi in the atmosphere of five climatologically diverse sites in the United States and Puerto Rico, *Curvularia* was the seventh most commonly cultured. It occurred at about the same frequency as *Epicoccum, Helminthosporium,* and *Aspergillus*, but this was only one-fourth the frequency of *Cladosporium* (Sorenson, Bulmer, and Criep 1974).

In 1945 Groves and Skolko treated 13 species of *Curvularia,* many of which they encountered as saprophytes in their investigations on seed-borne fungi. They illustrated most of the species by line drawings or photomicrographs. Ellis (1966) described and illustrated by line drawings 31 species distinguished on spore shape and the size and number of septations.

Erysiphe graminis DC. ex Mérat. stat. conid. *Oidium*   Mildew.

Ascomycetes   Erysiphales

Fructification.   The conidia or asexual spores appear on foliage of grasses as a whitish bloom. They are formed on short simple erect branches of the mycelium. As the season progresses the same mycelium forms cleistothecia, which are globoid and yellow changing to black, 150–250 μm in diameter and with a number of radiating irregular appendages. They contain numerous asci each with eight (rarely four) ascospores. It is the conidia that are usually found in aerial sampling.

Spore release.   The chains of conidia break apart readily and single spores or short chains become airborne. Conidia are formed and released throughout the growing season.

Substrate and distribution.   It is known throughout the world as an obligate parasite of cultivated and wild grasses.

Notes.   Conidia are usually found during the growing season. Hermansen et al. (1965) trapped them over Denmark at altitudes of 250-500 m and 1000 m during the period July 14 to August 5 in 1964. They found that the spores were more abundant at the lower altitudes and that the spore load decreased from the beginning to the end of the period, when conidium production gives way to the formation of cleistothecia. Their counts of trapped mildew conidia correspond well with the field observation of mildew severity of cereals in the same region. In a study of airborne molds in southern Ontario, Barlow (1963) found that conidia of *Oidium, Fusarium, Helminthosporium, Phoma, Rhizopus, Verticillium*, and others occurred rarely among plate-trapped spores.

Ascospores of *Erysiphe* are not known to the writer to be allergenic. In fact, few ascomycetes are recorded as being allergenic, but Frankland and

Gregory (1973) mentioned that *Leptosphaeria* (*Phoma*) has been so described. They also ascribe an asthmatic attack, during August, in a patient from the vicinity of Dorset, England, to be the result of exposure to ascospores of *Didymella exitialis* (Moreau) Müller. In August and September spores of this fungus were the most abundant of those trapped and were found in dense concentrations during periods of darkness. A local search disclosed that senescent leaves of barley and wheat bore numerous perithecia of *Didymella*. The patient's diary disclosed that the symptoms developed in parallel with the

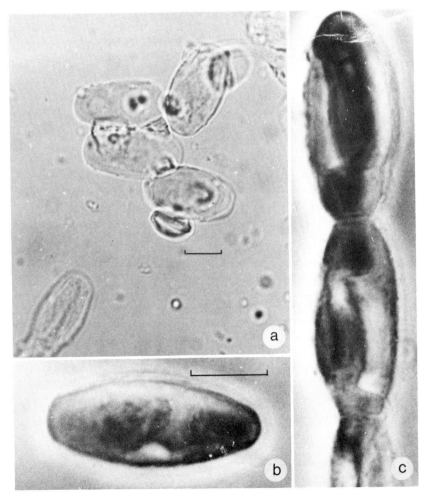

Fig. 149. *Erysiphe graminis* (stat. conid.). *a*, conidia, bright field; *b*, conidium, phase contrast; *c*, chain of conidia, phase contrast, magnification as in *b*. Scale lines = 10 μm.

ascospore concentration and declined in late August with the harvesting of the barley crop. Twelve of 100 patients with seasonal rhinitis reacted positively to extracts of the ascospores; *Didymella exitialis* is not known to occur in North America.

Ganoderma lucidum (Leyes. ex Fr.) Karst.   Conks, bracket fungi.

Basidiomycetes   Polyporaceae

Fructification. The conk or bracket occurs on trunks, stumps, or tree branches and is typically 7.5–30.5 cm in diameter, rich red brown, sessile, or with a lateral stalk. The underside is made up of whitish tubes lined with basidia, which bear basidiospores. Basidia also cover the gills of mushrooms and the teeth of hydnums, which also discharge basidiospores in great numbers.

Basidiospores. Ovoid, brown, 9–11 × 6–8 µm, slightly rough as seen under the light microscope, but smooth with internal markings when viewed through the scanning electron microscope (SEM).

Spore release. Spores are forcibly discharged from sterigmata at the apex of the basidium into the hollow of the tube and sift out into air currents. The fruit bodies, or conks, appear from spring through autumn but are tough and woody and may persist into winter. Wood-rotting fungi rarely release spores in winter, but airborne spores are sometimes trapped.

Substrate and distribution. It is found worldwide on coniferous and deciduous logs and stumps. In Canada it occurs commonly on *Tsuga* (hemlock).

Notes. This species is used as an example of the many fleshy, wood-rotting fungi that produce basidiospores. In addition to Polyporaceae (*Polyporus, Daedalea*) they include the Hydnaceae (*Hydnum*) and Thelephoraceae (*Thelephora, Tomentalla, Sistotrema*) and form basidiospores from pored, toothed, and effused hymenial surfaces respectively. The conks are conspicuous and their macroscopic characters, supported by microscopic characters, facilitate identification. However, the recognition of airborne basidiospores alone is a task for a specialist.

Reports of airborne fungus spores may include basidiospores, which might have more importance to the survey if they could be identified more precisely. Long and Kramer (1972) reported basidiospores trapped near woodland and prairie sites during conditions favorable (wet) and unfavorable (dry) for sporulation. Under wet conditions spore numbers peaked during the hours of darkness, but under dry conditions there was no peak period. The nighttime period at both sites corresponded to high relative humidity and lowered temperature. *Fomes* and *Poria,* both polypores, were reported by Lumpkins et al. (1973) in a correlated indoor and outdoor survey of airborne spores.

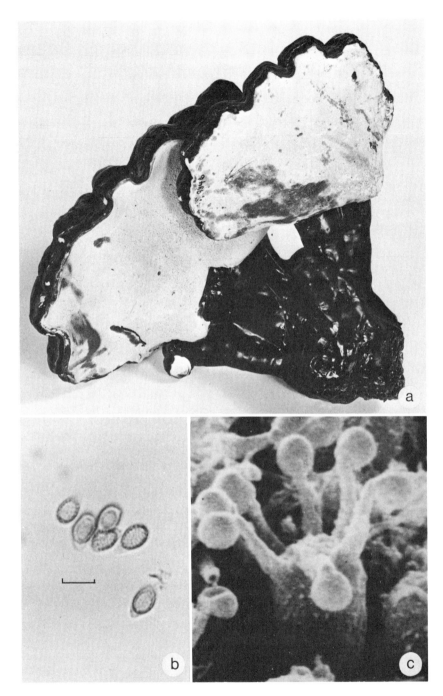

Fig. 150. *a–b, Ganoderma lucidum. a*, fruit body with pored undersurface, × ½; *b*, basidiospores, scale line = 10 μm. *c, Sistrotrema brinkmanii*, showing basidia and immature basidiospores, SEM × 9660, (Photo courtesy M. P. Corlett).

Fomes was about as common as *Penicillium*, but the counts were higher outdoors. *Poria* was decidedly less common and of the order of *Stemphlium* or *Curvularia*. The seasonal counts for *Fomes* varied, but in the eastern USA they were consistent throughout the year.

The wood-rotting fungi cause enormous damage to standing or fallen timber and processed lumber. They also play a role in reducing dead wood to soil nutrients.

Penicillium spp.

Fungi Imperfecti   Hyphomycetes

Fructification.   Aerial conidiophores, which eventually bear the spores, typically have a whisklike appearance and the sporulating colony is often green. The spores are produced from special flask-shaped cells known as phialides. They are supported by one or two rows of metulae that arise from the upper part of the conidiophore and not just from its apex.

Spores.   Catenate, ellipsoid or globose to lemon-shaped, smooth-walled or roughened, nonseptate, greatest dimension usually less than 10 μm, 3.7–9.4 × 2.5–4.4 μm in *P. digitatum* Sacc., 2.1–3.8 μm diam in *P. frequentans* Westling, and 2.5–3.0 × 2.0 μm in *P. lilacinum* Thom.

Spore release.   Chains of spores or single spores break away from the conidiophore presumably by the mechanical breaking of the primary cell wall that connects the spores. The incidence of *Penicillium* spores has been reported to peak in July and August, but other reports indicate seasonal independence (see notes).

Substrate and distribution.   Many species are soil-borne saprophytes, which occur throughout the world; some cause rot of plant matter including stored foods and animal products such as meat and cheese.

Notes.   Colonies of *Penicillium* are typically pale to dark green. Certain groups of species vary; *P. lilacinum* is pale lilac. In general, colony color darkens with maturity. Mutation may bring about noticeable color changes including various shades of red.

The genus *Penicillium* is divided into three main sections based on the characters of the spore-producing cells (phialides). These sections are further divided into series of species groups based on the characters of the colonies, the presence or absence of sclerotia and ascocarps (ascigerous or perfect state), the color of the colony, and spore morphology. The taxonomy has been thoroughly treated by Raper and Thom (1968), and an atlas depicting colony color with spore electron micrographs for representative species has been produced by Sakaguchi and Abe (1957).

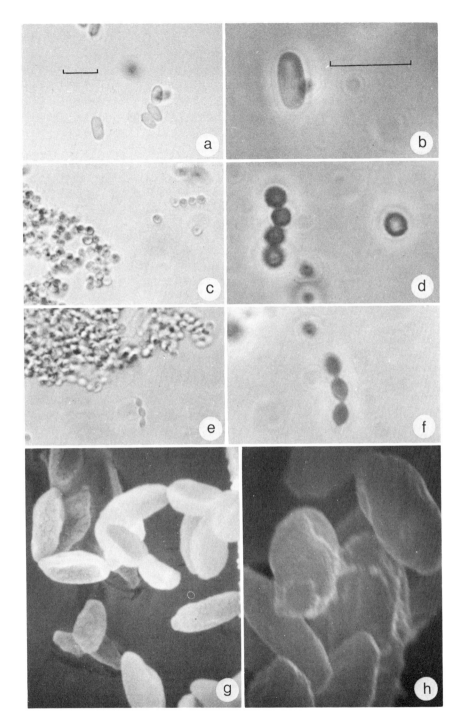

Fig. 151. *a–b*, *Penicillium digitatum*, conidia, scale lines = 10 μm. *c–d*, *P. frequentans*, conidia, magnification for *c* and *e* as in *a* and for *d* and *f* as in *b*. *e–h*, *P. lilacinum*, conidia; *g*, SEM × 7000; *h*, SEM × 16 800.

Fungi similar to *Penicillium* are *Gliocladium, Paecilomyces*, and *Scopulariopsis*. All have penicilliumlike conidiophores, but they are treated separately because they have, respectively, conidia in slime balls, unusually long phialides, and conidia truncate at the base with a basal pore and surrounding ring.

According to Raper and Thom (1968), *P. expansum* Link and *P. rubrum* Stoll are airborne molds that cause hay fever and asthmatic symptoms. Electron micrographs of the spores show each to be slightly roughened (?verrucose), rather elliptic in shape, and of similar size: $3.1-4.4 \times 2.5-3.1$ μm and $2.5-3.7 \times 1.5-3.4$ μm. Both species have been isolated from soil (Sakaguchi and Abe 1957) but are classified in different sections of the genus.

Walton and Dudley (1945) in their survey of airborne fungus spores in Manitoba collected *Penicillium* spores, but the numbers were not large enough to be of clinical importance. They concluded that spores of *Alternaria, Hormodendrum*, yeast, rusts, and smuts warranted attention as inhalent allergens. In samplings of air over the Atlantic Ocean, Pady and Kapica (1955) found *Penicillium* to comprise only 2% of the total spores collected. Collins-Williams et al. (1973) reported only a few spores of *Penicillium* in downtown Toronto, but Pady and Kapica (1956) found that they made up about 15% of the spores trapped in downtown Montreal and were present in the air throughout the year. Barlow (1963) found that *Penicillium* accounted for about 10% of the airborne spores in southern Ontario with a peak in July and August. The most abundant spores in the region were *Alternaria* and *Hormodendrum*. In a survey of home environments in the United States, Lumpkins et al. (1973) found that *Penicillium* was more abundant indoors, especially in basement rooms, than outdoors and was independent of the seasons or geography. Although spores were abundant, there was greater incidence of *Hormodendrum* (*Cladosporium*) and *Alternaria*.

Species of *Penicillium* are used for industrial purposes in the production of special cheeses, certain acids, and antibiotics.

Physarum cinereum (Batsch) Pers.   Slime mold.

Myxomycetes

Fructification.   Spores are formed in sporangia, which are sessile, gregarious, subglobose or elongate, and whitish with a purplish mass of spores within; this color is often visible as the sporangium breaks down.

Spores.   Purplish brown in mass, pale violaceous by transmitted light, globoid, 9-11 μm in diam; wall minutely echinate.

Spore release.   The sporangia are fragile when mature and spores are readily airborne when the substrate is disturbed. In the National Mycological Herbarium (DAOM), mature specimens bear collection dates from mid-May to early October.

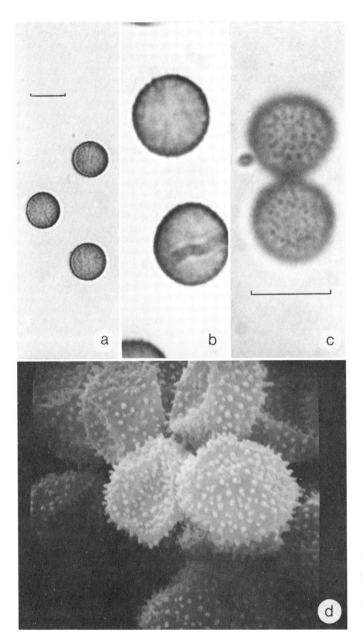

Fig. 152. *Physarum cinereum.* a, spores; b, spores in median plane, magnification as in c; c, spores in surface plane, scale lines = 10 μm; d, spores, SEM × 3640.

Substrate and distribution. It grows on living or dead plant matter at or just above ground level and frequently occurs on leaves and stems of grasses, especially lawn grasses. Some of the substrates named on specimens in DAOM include: red fescue, Kentucky bluegrass, alsike clover, decaying leaves, manure in greenhouses, dead wood, forest debris, and wintergreen.

Notes. Few of the 400 species of Myxomycetes have been found on slide traps, but their spores are present in the air and perhaps in abundance. They have been suggested by McElhenney and McGovern (1970) as a possible inhalent allergen along with actinomycetes and airborne algae. However, until Myxomycetes can be grown more readily in artificial culture or the spores alone can be more widely used in identifying the organisms, the assessment of Myxomycetes as allergens will remain a difficult task. One of the most recent taxonomic treatments of Myxomycetes (Martin and Alexopolus 1969) used characters of the fructification to separate orders, families, genera, and even some species (e.g. *Physarum*). *See* comparison plate of myxomycete spores and compare with *Ustilago*.

In some slime molds, spore liberation is greatly aided by the presence of a capillitium. This is a weft of slender strands, often elastic, which may expand to shed masses of spores to the air currents with the rupturing of the sporangium.

Puccinia graminis Pers.   Stem rust.

Basidiomycetes   Uredinales

Fructification. Spores form in orange, oblong, powdery sori called uredinia and push up from under the epidermis of grass stems and leaf sheaths often leaving the torn epidermis visible as a flap hinged at one side.

Urediniospores. Ellipsoid or oblong-ellipsoid, 20–42 × 13–24 μm; wall 1.5–2.0 μm thick, yellow brown, evenly, prominently echinate; germ pores (3–) 4, equatorial.

Spore release. Spores are formed singly on hyaline pedicels and easily break away at the point of attachment at the base of the spore to become airborne. When deposited on another part of the stem, another stem, or another plant, a spore germinates, incubates in the host, and develops a new sorus in about 10 days. The cycle repeats throughout the summer. Airborne rust inoculum has been monitored in Western Canada for many years by the Canada Department of Agriculture. In general airborne rust spores are rare during late May, increase slowly by late June, and then increase markedly, up to 100 times, by the end of August. The decline begins in late August (Walton and Dudley 1945; Green 1972).

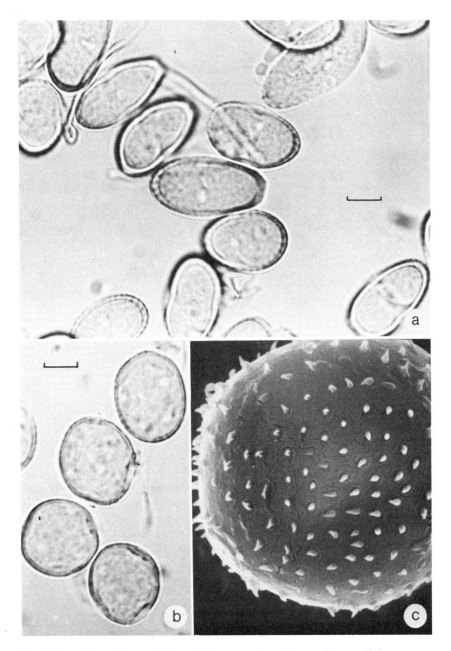

Fig. 153. *a, Puccinia graminis*, urediniospores. *b–c, P. recondita*, urediniospores; *c*, SEM × 3570; scale lines = 10 μm.

Habitat and distribution. The distribution is worldwide in temperate and tropical climates as an obligate parasite of grasses. Craigie (1957) demonstrated that stem rust urediniospores begin a northerly migration in the south central United States and arrive in the southern Canadian prairies in June and July. Hirst, Stedman, and Hurst (1967) reported unrediniospores high over the North Sea, probably originating east of the Baltic. Urediniospores are well adapted to aerial dissemination and in Canada they may be expected in abundance in the atmosphere during July and August.

Notes. Stem rust in Canada has caused severe loss to the wheat crop. Considerable effort is being made by the Canada Department of Agriculture and other agricultural agencies to reduce this loss and some success has been achieved.

Rusts have not been shown to be important as allergens, but Walton and Dudley (1945) concluded that in Manitoba rusts and smuts were of clinical importance. Urediniospores of stem rust would certainly be encountered in any summer survey of airborne pollen or fungus spores. They can be distinguished from most other rust urediniospores by their large size, strongly ellipsoidal shape, and conspicuous equatorial pores.

Puccinia recondita Rob. ex Desm. (*P. triticina* Erikss.)   Leaf rust.

Basidiomycetes   Uredinales

Fructification. Orange-colored, powdery uredinia are formed on both sides of the leaves of many grasses. The host epidermis ruptures usually by a central longitudinal slit. The sori are somewhat smaller than those of stem rust.

Urediniospores. Globoid or broadly ellipsoid, 20–34 × 20–24 μm; wall pale yellow brown, 1–2 μm thick, evenly and finely echinate; germ pores 6–10, scattered.

Spore release. In the past 10 years airborne leaf rust spores caught in spore traps in Western Canada exceeded stem rust spores by roughly 5–50 times; the numbers caught reach a maximum in August (Green 1972).

Habitat and distribution. Like *P. graminis*, this rust occurs wherever wheat is grown in Canada and occurs throughout the world on many other grasses.

Notes. Their globoid shape, small size, and scattered germ pores are the main characters that distinguish these urediniospores from those of *P. graminis*. Cummins (1971) listed 51 synonyms for this "species complex." The urediniospores closely resemble those of *P. coronata*, which attacks many grasses and cultivated oats.

The two-celled, dark brown teliospores of this species and *P. graminis* are important in the life cycles of these rusts, but they are not normally airborne and so are not described in detail here.

Stemphylium botryosum Wallr.

Fungi Imperfecti   Hyphomycetes

Fructification. Spores are formed from the tips of nodose or irregularly swollen, pale green brown, simple conidiophores, which grow singly or in small clumps.

Spores. Oblong, rounded at ends, 27–42 × 24–30 μm (teste Ellis, 1966), (20–) 24–33 (–35) × (12–) 15–24 (–26) μm (teste Simmons, 1967), pale to dark brown with usually 3 horizontal and 1–3 vertical septa (muriform), usually strongly constricted at the mid horizontal septum; wall smooth to verrucose.

Spore release. The spores are dry and readily become airborne.

Habitat and distribution. As a saprophyte on dead herbaceous plants it has worldwide distribution. Other species may occur as leaf-spotting parasites of specific hosts (e.g. *S. solani* on tomato and potato plants). It is frequently isolated from soil.

Notes. Certain species of *Stemphylium* and *Alternaria* have been relegated to the genus *Ulocladium* on the basis of geniculate rather than nodose conidiophores and ellipsoidal or pyriform rather than oblong or obpyriform spores. Generic recognition without cultured material requires critical observation. *Ulocladium consortiale* (Thum.) Simmons has been placed in *Stemphylium*, *Alternaria*, and *Pseudostemphylium* by competent authors at various times.

The key based on that by Ellis (1971) illustrates the conidial characters used in separating some common species of *Stemphylium.*

1. Spores rounded at apex
   2. Spores smooth (on *Trifolium*) .... ***S. sarciniforme***
   2. Spores verrucose (on dead herbaceous stems)
      3. Spores mostly constricted at median transverse septum.......... ***S. botryosum***
      3. Spores usually constricted at three major transverse septa .... ***S. vesicarium***

1. Spores with pointed conical apex

    4. Spores constricted at median transverse septum, length-to-breadth ratio not more than 2:1 .... ***S. solani***
    4. Spores constricted at three major transverse septa, length-to-breadth ratio 3:1 or more .... ***S. lycopersici***

Some species of *Stemphylium* are the imperfect states of ascomycetes; e.g. *S. botryosum* Wallr. is the conidial state of *Pleospora herbarum* (Fr.) Rabh. Simmons (1969) has demonstrated by cultures that at least five species have *Pleospora* as the ascigerous state.

Prince and Morrow (1969) include *Stemphylium* with other Dematiaceae (brown-spored Hyphomycetes) as giving the most striking reaction pattern in skin tests with patients in Texas. The particular reaction was in evidence in 56% of their mold-sensitive patients. If abundant, *Stemphylium* may be of importance as an allergen.

Pady and Kapica (1955, 1956) recorded small numbers of spores in studies of the air over the Atlantic Ocean and outdoors at Montreal. Lumpkins et al. (1973) made similar records indoors throughout the United States. Although usually present in aerial studies, *Stemphylium* is not recorded in the same abundance as *Alternaria* or *Cladosporium*.

Ustilago spp. Smut.

Basidiomycetes   Ustilaginales

Fructification. Spores form in sori or pustules on the stems, leaves, and inflorescences of host plants. Intercellular (rarely intracellular) mycelium aggregates into masses, and individual hyphae form intercalary spores in dark, often powdery masses.

Spores. One-celled, globoid, smooth to roughened, light to dark brown occasionally purple brown, often 10 μm or less but occasionally to 20 μm in diam.

Spore release. Wind or any mechanical action, such as threshing grain, disturbs the infected plants and dislodges the powdery spores to air currents.

Habitat and distribution. Smuts are found throughout the world; they parasitize many plant families, especially the grasses and cereals. Descriptions of smut species and their distribution in North America have been given by Fischer (1953). Other regional literature on smuts has been documented by Ainsworth (1971).

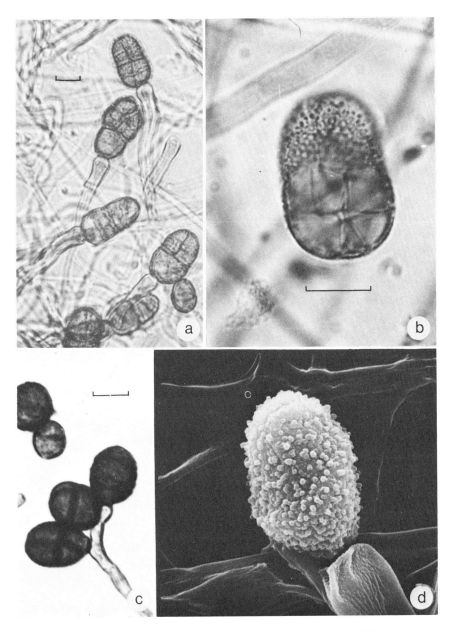

Fig. 154. *a–b, d, Stemphyllium botryosum. a–b,* conidia and conidiophores (Photo courtesy M. P. Corlett); *d,* young conidium, SEM × 3200. *c, Ulocladium ?atrum* Preuss (*Stemphylium consortiale* (Thuem.) Groves and Skolko). Scale lines = 10 μm.

Notes. *Ustilago avenae* (Pers.) Rostr. on *Avena* (oats) and *Hordeum* (barley) infects the inflorescence and replaces the seed by a black, powdery mass of spores. The spores are globose, yellowish brown, 6–9 × 5–7 μm; minute roughenings on the walls are more conspicuous on one side when viewed by the light microscope. The wall sculpturing is plainly verrucose when seen by the scanning electron microscope.

Ustilago hordei (Pers.) Lagerh. on the same and other hosts has spores with smooth walls, and the infected spikelets are blackened but with an agglutinated and not a powdery spore mass.

Ustilago maydis (DC.) Cda. (*U. zeae* Ung.) appears on the stems, leaves, and inflorescences, especially on corn (*Zea mays* L.), as small or large galls made of black powdery masses. The spores are globose, light olive brown, 7–10 μm in diameter and prominently echinate over the entire surface when viewed under the light microscope.

Spore sampling methods using culture plates give little information about smuts or rusts. These fungi, especially the smuts, can be grown on artificial media, but special techniques and media are required. More information can be obtained on airborne smut and rust spores by the use of microscope slide traps. In a survey of the airborne fungus spores in Manitoba from 1939 to 1944 Walton and Dudley (1945) found many fungus colonies on agar. The most common were *Hormodendrum* and *Alternaria*, but there were no rusts or smuts. Using traps with microscope slides, average spore counts of rusts (July and August) and of smuts (July) exceeded the spore counts of *Hormodendrum* and *Alternaria* for these same peak months. The study concluded that in Manitoba rusts and smuts along with *Alternaria, Hormodendrum, Monilia,* and yeasts were clinically important and should be considered in the treatment of inhalant allergies.

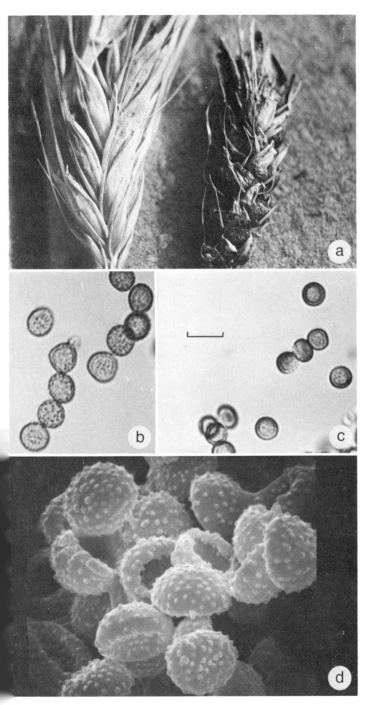

Fig. 155. *Ustilago* spp. *a*, *U. hordei*, healthy (*left*) and infected (*right*) spikes of *Hordeum* (barley), × 1. *b*, *U. maydis*, spores, magnification as in c. *c–d*, *U. avenae*, spores; *d*, SEM × 3640; scale line = 10 μm.

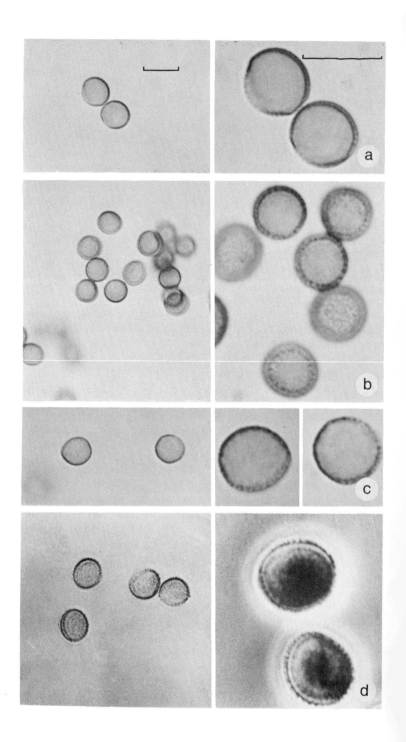

Fig. 156. Comparison plate of Myxomycetes, magnifications in left column as in *a* left and in right column as in *a* right, scale lines = 10 μm. a, *Comatricha nigra*, spores globose, about 9–10 μm in diam; black to purplish in mass, dark violet by transmitted light; wall faintly warted. b, *Fuligo septica*, spores spherical, about 6–9 μm in diam; dull black in mass, purplish brown by transmitted light; wall minutely spinulose. c, *Didymium megalsporum* (*D. eximium*), spores globose, about 8–10 μm in diam; black in mass, violet brown by transmitted light; wall minutely warted. d, *Metatrichea vesparium* (*Hemitrichia v.*), spores globose, about 9–11 μm in diam; brownish red in mass, yellow by transmitted light; wall minutely spined, (right in phase contrast).

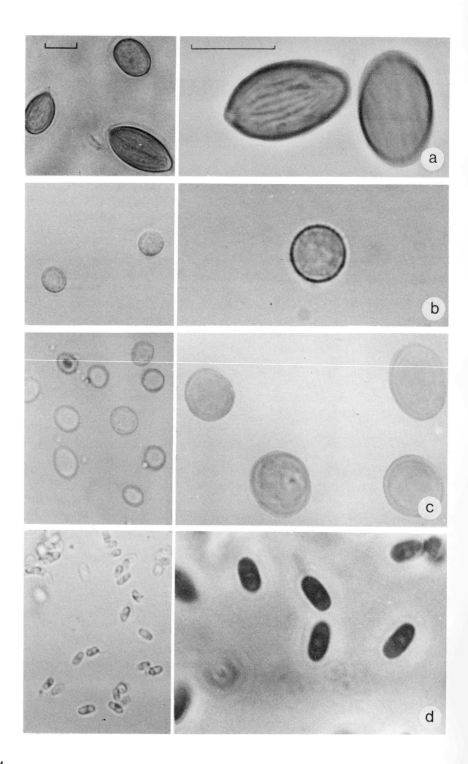

Fig. 157. Comparison plate of Phycomycetes, magnifications in left column as in *a* left and in right column as in *a* right, scale lines = 10 μm. *a, Choanephora cucurbitarum*, conidia ellipsoidal, slightly protruding hilum; wall very shallowly rugose, pale yellow brown (right in phase contrast). *b, Cunninghamella elegans*, conidia spherical; wall faintly echinate, pale yellow. *c, Rhizopus nigricans*, conidia globoid; wall smooth, hyaline. *d, Mucor ambiguus*, conidia oblong-bacillar, mostly biguttulate; wall smooth, hyaline, pale yellow, (right in phase contrast).

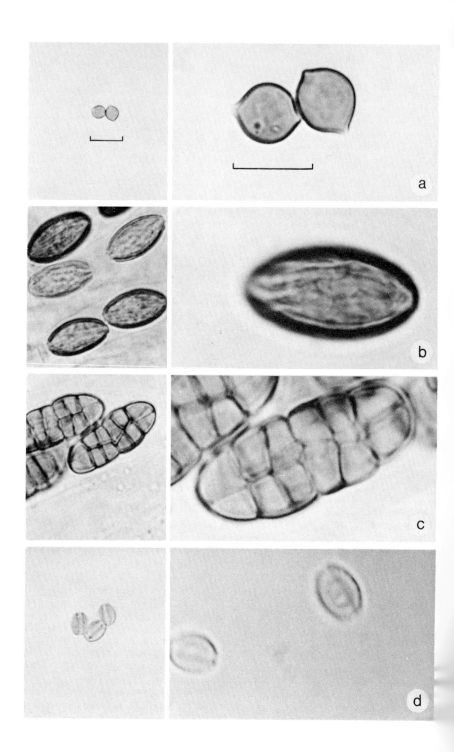

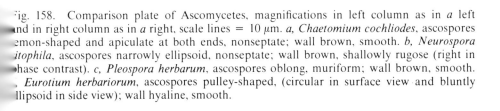

Fig. 158. Comparison plate of Ascomycetes, magnifications in left column as in *a* left and in right column as in *a* right, scale lines = 10 μm. *a, Chaetomium cochliodes*, ascospores lemon-shaped and apiculate at both ends, nonseptate; wall brown, smooth. *b, Neurospora sitophila*, ascospores narrowly ellipsoid, nonseptate; wall brown, shallowly rugose (right in phase contrast). *c, Pleospora herbarum*, ascospores oblong, muriform; wall brown, smooth. *d, Eurotium herbariorum*, ascospores pulley-shaped, (circular in surface view and bluntly ellipsoid in side view); wall hyaline, smooth.

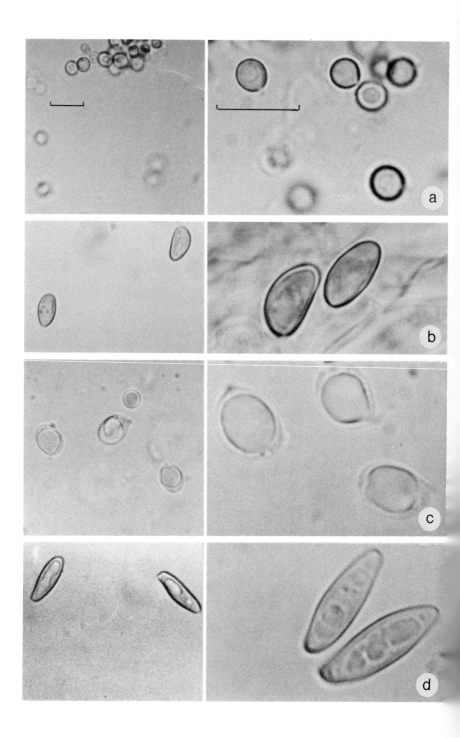

Fig. 159. Comparison plate of Basidiomycetes, magnifications in left column as in *a* left and in right column as in *a* right, scale lines = 10 μm. *a, Calvatia gigantea*, basidiospores spherical, apiculus peglike; wall smooth, pale yellow. *b, Serpula lacrymans*, basidiospores unequally ellipsoid; wall smooth, bright yellow. *c, Amanita muscaria*, basidiospores globoid to broadly ellipsoid; wall smooth, hyaline. *d, Boletus edulis*, basidiospores narrowly ellipsoid, guttulate; wall smooth, pale yellow brown.

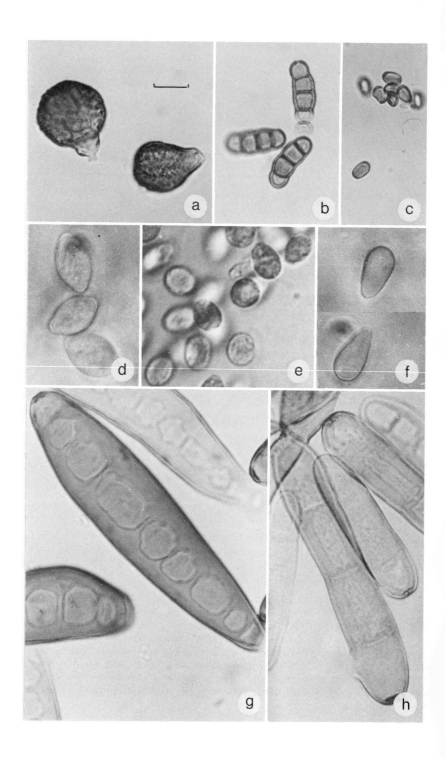

Fig. 160. Comparison plate of Fungi Imperfecti, magnifications as in a, scale line = 10 μm. a, *Epicoccum purpurascens*, conidia subspherical to pyriform, muriform, brown with paler protuberant stalk cell; wall verrucose. b, *Torula herbarum*, conidia cylindrical to oblong, straight to slightly curved, mostly 3–4 horizontally septate; wall minutely verrucose. c, *Cryptostroma corticale*, conidia ovoid to oblong, brown, nonseptate; wall smooth. d, *Monilia* sp. (perf. stat. *Monilinia fructicola*) conidia lemon-shaped, hyaline; wall smooth. e, *Phyllosticta minima*, conidia broadly ellipsoid, hyaline, nonseptate; wall smooth. f, *Botrytis cinerea* (perf. stat. *Botryotinia fuckeliana*), conidia ovoid to obovoid, hyaline, nonseptate; wall smooth. g, *Bipolaris sorokiniana* (*Helminthosporium sativum*; perf. stat. *Cochliobolus*), conidia narrowly ellipsoid to obclavate, many horizontal pseudosepta; wall smooth, brown; conidia produced from lateral pores in the cylindrical conidiophores. h, *Drechslera teres* (perf. stat. *Pyrenophora*), conidia cylindrical, fusiform or obclavate, straight or slightly curved, many horizontal pseudosepta; wall smooth, brown; conidia formed sympodially on conidiophores.

GLOSSARY

apothecium (*pl.* apothecia)   A cuplike structure, sessile or stalked, bearing asci on the upper surface.
ascocarp   A structure housing asci, often specialized in dehiscence e.g. apothecium, cleistothecium, hysterothecium, perithecium.
ascus (*pl.* asci)   A specialized cell containing reproductive sexual units called ascospores.
basidiocarp   A structure bearing basidia usually from gilled, pored, toothed, or undifferentiated surfaces.
basidium (*pl.* basidia)   A specialized cell bearing reproductive sexual units called basidiospores.
catenate   Spores produced in chains.
cleistothecium (*pl.* cleistothecia)   A closed structure containing asci within and with no preformed region of dehiscence.
conidiogenous cell   Any cell that produces a conidium.
conidiophore   A conidiogenous cell or cells plus supporting structures.
conidium (*pl.* conidia)   Any asexual spore except a sporangiospore or intercalary spore.
echinulate   Sharp pointed warts, cf. verrucose.
hilum   The scar on a spore indicating the attachment point to a conidiophore or pedicel.
hypha (*pl.* hyphae)   A slender, threadlike chain of vegetative cells.
hysterothecium   A closed structure bearing asci and dehiscing along a preformed slit.
metula (*pl.* metulae)   A special hyphal cell that bears the spore-producing phialide.
muriform   Having both horizontal and vertical septa as in *Alternaria* spores.
perithecium   A closed structure bearing asci and opening by a pore; the pore may be raised up into a beak or stretched into a neck.
phialide   A hyphal cell with an open end or a succession of open ends from which conidia emerge.
septum (*pl.* septa)   A cross wall in hyphae or spores.
sporangium (*pl.* sporangia)   An organ that produces asexual spores endogenously; cf. basidium and ascus.
spore   A general term for a reproductive unit that in the fungi may have one or more cells.
sporogenous hypha   A specialized hypha that produces the reproductive units or spores, e.g. a conidiophore, a phialide.
sterigma (*pl.* sterigmata)   A protrusion from a cell used to support a spore, especially on the basidium; the term is used synonymously with metula in some older works on *Penicillium*.
sympodial   Spores are produced at the apex of a conidiophore, and when growth resumes from below the spore, the conidiophore has a geniculate (kneed) appearance (see *Curvularia*).
toxin   A chemical compound that acts as a poison.

urediniospore A rust spore, usually one-celled, echinate and borne singly; responsible for rapid buildup of the fungus during periods favorable for the growth of the fungus.

uredinium (*pl.* uredinia) A fructification in rust fungi that arises on a dikaryotic mycelium and gives rise to repeating spores; other spellings include uredium and uredosorus.

verrucose Round topped warts.

REFERENCES

Ainsworth, G. C. 1971. Ainsworth and Bisby's dictionary of the fungi. 6th ed. Commonw. Mycol. Inst., Kew, Surrey, England. 631 pp.

Arrhenius, E. 1973. Mycotoxicosis — an old health hazard with new dimensions. Ambio 2:49-56.

Austwick, P. K. C. 1965. Pathogenicity. Pages 82-126 in K. B. Raper and D. I. Fennell, The genus *Aspergillus*. Williams and Wilkins Co., Baltimore, Md.

Barlow, H. S. 1963. A collaborative study of airborne molds in southern and central Ontario. Ann. Allergy 21:569-576.

Barron, G. L. 1968. The genera of Hyphomycetes from soil. Williams and Wilkins Co., Baltimore, Md. 364 pp.

Chaterjee, J., and Hargreave, F. E. 1973. Atmospheric pollen and fungal spores in Hamilton, Ontario, 1972. J. Allergy Clin. Immunol. 50:124. (Abstr.)

Christensen, C. M. 1965. The molds and man. An introduction to the fungi. 3rd ed. revised. University of Minnesota Press, Minneapolis, Minn. 284 pp.

Collins-Williams, C., Nizami, R. M., Lamenza, C., and Chiu, A. W. 1972. Nasal provocative testing with molds in the diagnosis of perennial allergic rhinitis. Ann. Allergy 30:557-561.

Collins-Williams, C., Kuo, H. K., Garey, D. N., Davidson, S., Collins-Williams, D., Fitch, M., and Fischer, J. B. 1973. Atmospheric mold counts in Toronto, Canada, 1971. Ann. Allergy 31:69-71.

Collins-Williams, C., Kuo, H. K., Langer, H., Doron, I. G., Lovera, J., and Baboo, M. 1973. Provocative bronchial testing with molds. Ann. Allergy 31:401-406.

Cooke, W. B. 1959. An ecological life history of *Aureobasidium pullulans* (de Bary) Arnaud. Mycopathol. Mycol. Appl. 12:1-49.

Craigie, J. H. 1957. Stem rust of cereals. Can. Dep. Agric. Publ. 666. 45 pp.

Cummins, G. B. 1971. The rust fungi of cereals, grasses and bamboos. Springer-Verlag New York Inc., New York, N.Y. 569 pp.

Ellis, M. B. 1966. Dematiaceous Hyphomycetes. VII. *Curvularia, Brachysporium*. Mycological Papers 106. Commonw. Mycol. Inst., Kew, Surrey, England. 55 pp.

Ellis, M. B. 1971. Dematiaceous Hyphomycetes. Commonw. Mycol. Inst., Kew, Surrey, England. 608 pp.

Fischer, G. W. 1953. Manual of the North American smut fungi. Ronald Press Co., New York, N.Y. 343 pp.

Frankland, A. W., and Gregory, P. H. 1973. Allergenic and agricultural implications of airborne ascospore concentrations from a fungus *Didymella exitialis*. Nature (Lond.) 245:336-337.

Green, G. J. 1972. Airborne rust inoculum over Western Canada in 1971. Can. Plant Dis. Surv. 52:6-7.

Gregory, P. H. 1961. The microbiology of the atmosphere. Leonard Hill [Books] Ltd., London, England. 251 pp.

Groves, J. W., and Skolko, A. J. 1945. Notes on seed-borne fungi. III. *Curvularia*. Can. J. Res. (C) 23:94-104.

Hermansen, J. E., Johansen, H. B., Hansen, H. W., and Carstensen, P. 1965. Notes on the trapping of powdery mildew conidia and urediospores by aircraft in Denmark in 1964. Pages 121–129 *in* R. Vet. and Agric. Coll. Yearbook 1965. Copenhagen, Denmark.

Hirst, J. M., Stedman, O. J., and Hurst, G. W. 1967. Long distance spore transport: Vertical sections of spore clouds over the sea. J. Gen. Microbiol. 48:357–377.

Hughes, S. J. 1958. Revisiones Hyphomycetum aliquot cum appendice de nominibus rejiciendis. Can. J. Bot. 36:727–836.

Hyde, H. A. 1972. Atmospheric pollen and spores in relation to allergy. I. Clin. Allergy 2:153–179.

Ingold, C. T. 1971. Fungal spores: Their liberation and dispersal. Clarendon Press, Oxford, England. 302 pp.

Kendrick, W. B., editor. 1971. Taxonomy of Fungi Imperfecti. University of Toronto Press, Toronto, Ont. 309 pp.

Long, D. L., and Kramer, C. L. 1972. Air spora of two contrasting ecological sites in Kansas. J. Allergy Clin. Immunol. 49:255–266.

Lumpkins, E. D., Sr., Corbit, S. L., and Tiedeman, G. M. 1973. Airborne fungi survey. I. Culture plate survey of the home environment. Ann. Allergy 31:361–370.

Martin, J. W., and Alexopolus, C. J. 1969. The Myxomycetes. University of Iowa Press, Iowa City, Iowa. 560 pp.

McElhenney, J. T., and McGovern, J. P. 1970. Possible new inhalent allergens. Ann. Allergy 28:467–471.

Mishra, R. R., and Srivastava, V. B. 1972. Aeromycology of Gorakhpur. V. Airspora over wheat and barley fields. Mycopathol. Mycol. Appl. 47:349–355.

Pady, S. M., and Kapica, L. 1955. Fungi in the air over the Atlantic Ocean. Mycologia 47:34–50.

Pady, S. M., and Kapica, L. 1956. Fungi in air masses over Montreal during 1950 and 1951. Can. J. Bot. 34:1–15.

Pady, S. M., Kramer, C. L., and Clary, R. 1969. Periodicity in spore release in *Cladosporium.* Mycologia 61:87–98.

Parmelee, J. A. 1967. The autoecious species of *Puccinia* on Heliantheae in North America. Can. J. Bot. 45:2267–2327.

Prince, H. E., and Morrow, M. B. 1969. A logical approach to mold allergy. Ann. Allergy 27:79–86.

Raper, K. B., and Thom, C. 1968. A manual of the *Penicillia.* Hafner Publishing Co., New York, N.Y. 875 pp.

Raper, K. B., and Fennell, D. I. 1965. The genus *Aspergillus.* Williams and Wilkins Co., Baltimore, Md. 686 pp.

Sakaguchi, Kin-ichiro, and Shigro Abe. 1957. Atlas of micro-organisms. The *Penicillia.* Kanehara Shuppan Co. Ltd., Tokoyo, Japan. 319 pp.

Schlueter, D., Fink, J., and Hensley, G. 1973. Woodpulp workers' disease: a hypersensitivity pneumonitis caused by *Alternaria.* J. Allergy Clin. Immunol. 50:36 (Abstr.)

Simmons, E. G. 1967. Typification of *Alternaria, Stemphylium* and *Ulocladium.* Mycologia 59:67–92.

Simmons, E. G. 1969. Perfect states of *Stemphylium.* Mycologia 61:1–26.

Sorenson, W. G., Bulmer, G. S., and Criep, L. H. 1974. Airborne fungi from five sites in continental United States and Puerto Rico. Ann. Allergy 33:131–137.

Unligil, H. H., Shih, M. S. H., and Shields, J. K. 1974. Airborne fungal spores at lumber seasoning yards in the lower Ottawa Valley. Can. J. For. Res. 4:301–307.

Walton, C. H. A., and Dudley, Margaret G. 1945. Airborne fungus spores in Manitoba. Can. Med. Assoc. J. 53:529–538.

Widden, P., and Parkinson, D. 1973. Fungi from Canadian forest soils. Can. J. Bot. 51:2275–2290.

ACKNOWLEDGMENTS

We wish to express appreciation to Dr. J. Terasmae, Department of Geological Sciences, Brock University, St. Catharines, Ontario, and colleagues in the Biosystematics Research Institute for their advice, constructive criticism, and reviews of the manuscript. The scanning electron microscope facilities were made available through the kindness of the staff of the Chemistry and Biology Research Institute. We are grateful for the assistance provided by the staff of the Graphics Section and C. R. Wood, Scientific Editor, Research Program Service.

INDEX OF SPECIES DESCRIBED

VASCULAR PLANTS

Abies balsamea, 58
Abies grandis, 60
Abies lasiocarpa, 60
Acer circinatum, 97
Acer glabrum var. *douglasii,* 98
Acer macrophyllum, 100
Acer negundo, 101
Acer nigrum, 102
Acer pensylvanicum, 104
Acer rubrum, 106
Acer saccharinum, 107
Acer saccharum, 109
Acer spicatum, 111
Agropyron repens, 192
Alnus crispa, 115
Alnus rugosa, 117
Amaranthus albus, 113
Amaranthus retroflexus, 114
Ambrosia acanthicarpa, 146
Ambrosia artemisiifolia, 147
Ambrosia chamissonis, 149
Ambrosia psilostachya, 151
Ambrosia trifida, 152
Arenaria serpyllifolia, 132
Artemisia biennis, 163
Artemisia campestris, 164
Artemisia frigida, 165
Artemisia ludoviciana, 167
Artemisia tridentata, 169
Artemisia vulgaris, 170
Atriplex subspicata, 135
Axyris amaranthoides, 133

Betula alleghaniensis, 118
Betula minor, 120
Betula occidentalis, 121
Betula papyrifera, 122
Betula populifolia, 124
Boehmeria cylindrica, 253

Cannabis sativa, 128
Cardamine pratensis, 171
Carex aquatilis, 173

Carya cordiformis, 199
Carya glabra, 201
Carya ovata, 201
Castanea dentata, 177
Celtis occidentalis, 251
Chamaecyparis nootkatensis, 50
Chenopodium album, 136
Chenopodium berlandieri var. *zschackei,* 137
Comptonia peregrina, 210
Corylus cornuta, 125

Dactylis glomerata, 193

Equisetum arvense, 48
Eriogonum flavum, 228
Eurotia lanata, 139

Fagus grandifolia, 178
Fraxinus americana, 212
Fraxinus nigra, 213
Fraxinus pennsylvanica, 214

Ginkgo biloba, 55

Iva axillaris, 154
Iva frutescens, 156
Iva xanthifolia, 157

Juglans cinerea, 202
Juglans nigra, 204
Juniperus communis, 52

Kochia scoparia, 140

Laportea canadensis, 254
Larix laricina, 61
Larix lyallii, 63
Larix occidentalis, 63
Littorella americana, 216
Luzula multiflora, 204
Lycopodium selago, 49

Morus rubra, 207

Myrica gale, 208
Myrica pensylvanica, 209

Ostrya virginiana, 126
Oxyria digyna, 229

Parietaria pensylvanica, 255
Phleum pratense, 195
Picea glauca, 64
Picea mariana, 66
Picea rubens, 68
Pilea pumila, 256
Pinus albicaulis, 70
Pinus banksiana, 72
Pinus contorta, 74
Pinus flexilis, 76
Pinus monticola, 78
Pinus ponderosa, 80
Pinus resinosa, 81
Pinus rigida, 82
Pinus strobus, 85
Plantago canescens, 218
Plantago eriopoda, 219
Plantago lanceolata, 220
Plantago macrocarpa, 221
Plantago major, 222
Plantago maritima, 223
Plantago rugelii, 225
Platanus occidentalis, 226
Poa pratensis, 196
Polygonum lapathifolium, 231
Populus grandidentata, 241
Pseudotsuga menziesii, 87

Quercus alba, 180
Quercus bicolor, 182
Quercus garryana, 183
Quercus macrocarpa, 184
Quercus muehlenbergii, 185
Quercus palustris, 187
Quercus rubra, 189
Quercus velutina, 190

Rumex acetosa, 233
Rumex acetosella, 235
Rumex obtusifolius, 236
Rumex orbiculatus, 237

Salix discolor, 243
Salsola pestifer, 141
Sambucus canadensis, 130
Sanguisorba canadensis, 240
Sarcobatus vermiculatus, 143
Shepherdia argentea, 174
Shepherdia canadensis, 175
Solidago canadensis, 158
Suaeda maritima, 144

Taxus brevifolia, 94
Thalictrum dasycarpum, 238
Thuja occidentalis, 54
Thuja plicata, 55
Tilia americana, 244
Tsuga canadensis, 89
Tsuga heterophylla, 91
Tsuga mertensiana, 93
Triglochin maritima, 205
Typha angustifolia, 245
Typha latifolia, 247

Ulmus americana, 249
Ulmus rubra, 250
Urtica dioica ssp. *gracilis,* 257

Xanthium spinosum, 160
Xanthium strumarium, 161

Zea mays, 198

FUNGI

Alternaria alternata, 275
Amanita muscaria, 309
Aspergillus flavus, 277
Aspergillus niger, 277
Aureobasidium pullulans, 279

Bipolaris sorokiniana, 311
Boletus edulis, 309
Botrytis cinerea, 311

Calvatia gigantea, 309
Chaetomium cochliodes, 307
Choanephora cucurbitarum, 305
Cladosporium herbarum, 281

Comatricha nigra, 303
Crypostroma corticale, 311
Cunninghamella elegans, 305
Curvularia spp., 284

Didymium megalosporum, 302
Drechslera teres, 311

Epicoccum purpurascens, 311
Erysiphe graminis, 286
Eurotium herbariorum, 307

Fuligo septica, 303

Ganoderma lucidum, 288

Metatrichea vesparium, 303
Monilia sp., 311
Mucor ambiguus, 304

Neurospora sitophila, 307

Penicillium spp., 290
Phyllosticta minima, 311
Physarum cinereum, 292
Pleospora herbarum, 307
Puccinia graminis, 294
Puccinia recondita, 296

Rhizopus nigricans, 304

Serpula lacrymans, 309
Stemphyllium botryosum, 297

Torula herbarum, 311

Ustilago spp., 298